The Student Guide to

BTEC

Business and Finance

Reg Chapman
Oldham College

Alistair Norman
Burnley College

Marie Norman
Bolton Institute

Series Editor: **Alan Kitson**
Bolton Institute

MACMILLAN

First edition 1991
Reprinted once
Second edition 1993
Published by
THE MACMILLAN PRESS LTD
Houndmills, Basingstoke, Hampshire RG21 2XS
and London
Companies and representatives
throughout the world

ISBN 0–333–58687–5

A catalogue record for this book is available
from the British Library

Printed in Hong Kong

10 9 8 7 6 5 4 3
00 99 98 97 96 95 94 93

Contents

UNIT 1 | Introduction

In this unit, you will:

- learn what this book can do for you;

- learn how the book is split up into units and what topics are covered in each unit;

- discover how you should make use of the book and, especially, the many activities in each unit.

A job with prospects

If you are reading this book, then you already have a commitment to a career in business or are actively considering such a career. Not just any old job will do – you want a job with prospects. You may already have a job and want to improve your chances of promotion, or you may be a full-time student looking to get the first step in a career in business. Whether you are a full- or part-time student, you are one of a large and growing number of people who are seeking to achieve their career ambitions by taking a BTEC National in Business and Finance.

So, what is 'business'? Business doesn't just mean big business and it doesn't just mean firms that make money. Business includes: the Government – local councils and the Civil Service; the voluntary sector – organisations like Oxfam, Greenpeace, Age Concern and Friends of the Earth; small firms – local shops, solicitors, accountants or manufacturers; and big firms with branches across the country or across the world – IBM, ICI or Philips. Business includes all of the organisations that need to control their work efficiently and effectively; and that means a lot of organisations need people with a knowledge of how to get on in business. This may be a broader way of looking at business than you are used to, but it's one that is used throughout the book, as it reflects the way that BTEC programmes are delivered.

Activity 1   Just arrived

You have just started on a BTEC National course or programme (the two terms mean the same thing) and you want to be successful at college, whether full- or part-time. This means that you need to know what you have to put into the course and what you want to get out of it.

Use Table 1 to list what you will have to do to be a successful student – that is, what you have to put into the course. Then list what you hope to get out of the course. Be honest with yourself.

First of all, you should fill in the table yourself. Then compare your list with that of someone else. If you do this exercise in class, it is valuable for each person or pair to explain their lists to the whole group. The similarities and differences between lists make for some useful discussion and will help you to set your stall out to be successful on the course.

Some suggestions have been included in the table to get you started – but you may not agree with them.

TABLE 1   *Your success factors*

What I will have to put into the course	What I want to get out of the course
Attend classes regularly	A recognised qualification
Work with other students	Sample different jobs in business
Do a lot of course work	Develop skill with computers

This book will help you to get the most out of your BTEC programme and the best possible grades at the end. The book contains:

- All the basic information you need on the content and organisation of the course.
- A detailed and practical explanation of the BTEC style of teaching, learning and assessment.
- Lots of examples of the different types of assignment used in your course and tips on how to tackle them.
- Advice on where to get information on business while you are on the course.
- Ideas on what you might do after the course.

The book is written as a practical handbook for you to work through (perhaps with your tutor). It is a *workbook*: a mixture of information that you need to know and practical tasks or projects that you can tackle on your own or with others on the course. You will find it especially useful for the first few months of the course, when everything seems strange to you. At that stage, you will probably have lots of questions that you want answering, and this book aims to answer most of them for you, or tells you how to go about finding the answer for yourself. However, the book isn't just for the first few months, as you will find lots of assignments to practise on throughout the course, especially the first year. It also gives you tips on how to approach work, which you can refer back to at any time through the course, as well as telling you about things that will crop up after the first few months, such as work experience or final assignments. It provides some useful discussion topics for your tutorials – that is, the time set aside for discussion of any problems with the lecturer responsible for your class.

Organisation of the book

The book is divided into eight units – this introductory unit and seven more dealing with all of the main aspects of the course in more detail. The units fall into three groups: Units 1 to 3, Units 4 to 7 and Unit 8.

Units 1 to 3

These units give you a clear outline of most of the main aspects of your BTEC programme and answer some of your first questions.

- Unit 1, *Introduction*, deals with questions like:
  - What is the book for?
  - How should I work through the book?

- Unit 2, *Learning with BTEC*, deals with questions like:
  - What is BTEC?
  - What qualifications does BTEC offer?
  - Why should I choose a BTEC programme in Business and Finance?
  - What will I do on the course?
  - What can I do after the course?

- Unit 3, *Learning Business and Finance with BTEC*, deals with questions like:
  - What subjects will I learn?
  - How will I be taught?
  - Who can provide help and advice?
  - How can I improve my work and get the most out of the course?

Units 4 to 7

These units deal with the main area that students worry about on a BTEC National – assessment and grading. Because this is so important to you, four units are devoted to this area.

- Unit 4, *Tackling Assignments*, is the longest unit in the book and covers questions like:
  - What are assignments?
  - What sorts of assignment will I have to do?
  - How do I tackle the different tasks?

- Unit 5, *Assessment*, goes into considerable detail to answer all of your questions about BTEC's very distinctive style of assessment and grading. It deals with questions like:
  - How am I going to be assessed?
  - What are assessment criteria?
  - How do I show what I can do?

- Unit 6, *Core Assignments to Practise On*, and Unit 7, *Assignment Programmes*, are quite different from previous units. Unit 6 consists of lots of different individual assignments for you to tackle and Unit 7 shows how they all fit together into a programme of assignments. Tips on how to go about each assignment are provided to help you.

Unit 8

Sources of Help and Information is the final unit, but it is one that you can use from the very start of the course. It provides you with advice on how and where to get information on business organisations and events – something that you will often have to do. It helps you with questions like:

- Where can I find information on ...?
- Who can help me with grants?
- Where do I go to get a copy of employment law pamphlets?
- What facilities are there in college for me to use?

Getting information on BTEC

If you aren't sure whether a BTEC National is right for you, then you need to collect background information on the course and get advice from your local BTEC centre, where the course will be taught. You should:

- Get in touch with the co-ordinator for the BTEC National in the Business Studies section of the centre. This is the lecturer in overall charge of the course, who will be able to send you information on the course and make an appointment for you to go in to talk about the course and have a look around. Don't be put off if you get an application form to fill in with the information that you receive. You don't have to make up your mind at this stage, and filling in the form merely gives the centre an idea of the numbers of those who might be interested in the course, as well as providing the staff with an idea of your background. This will help them to advise you as to whether the course is suitable for you.
- Before you go to the centre to discuss whether the course is right for you, make a list of the questions that you want to ask. Table 2 gives some of the main ones.

TABLE 2   *Questions that you might ask*

Question	Answer
How much does the course cost?	
How do I get to college?	
Can I get a bus pass?	
What did previous students go on to do?	
What days and times will I be in college?	
What subjects can I take?	

- Get as much independent advice as you can about the course and the centre – ask students that you know, or check with the careers service or careers teachers at school. If you're in a job, try to find someone who has done the course, or ask your supervisor or line manager what he/she thinks; the training officer or personnel section may also be able to advise you.

Unit sequence

You don't have to read through the units in the order given. If, for example, you need to know about assessment immediately, then you could go straight to Unit 5. You'll get more out of the book though if you work your way through Units 2 and 3 before going on to the more detailed units.

Unit layout

Each unit has a common layout:

• Unit aims, which state briefly, at the start of the unit, what you will get out of it.
• Text, which contains the information that you need.
• Activities, which you work through to help you to learn. They are set within the text of each unit and you will usually find some tips on how to go about the activity, as well as some suggested answers. Often, there is no single 'right' answer, so your answers may differ slightly from the ones that are suggested.

Activities

This book is a workbook – you are not meant just to read it from cover to cover. You have to work through the activities with the guidance of your tutor to find things out, present information or do a short project. This means that you learn from doing the activities, as much as from the information that you get from the text.

Activity 2   Ice-breaker

At the start of the course, you need to get to know the other students and get over any shyness that you might feel in speaking to your class. This activity breaks the ice in your class and starts you thinking about yourself and your career. Other activities in Units 2 and 3 will help you to do a personal stock-take – to build a picture of yourself, your strengths and weaknesses, and career plans.

1   Join with another member of your class who you don't know. Make sure there is a space between you and the rest of the class.

2   Take it in turns to find out all that you can about one another for five minutes each. You might ask about age, name, schools, jobs, reasons for coming on the course, hobbies and interests, and career plans. Make a note of the answers and write them on a flip chart.

3   At the end of 10 minutes, each person should speak for two to three minutes to the whole class. Tell them all that you have found out about the person that you were paired with. Give the person you have been talking about a chance to correct any mistakes or add anything that he/she feels is important.

4   The details of all of the people in the class should be put on to one form or flip chart, so that everyone can take a copy.

5   Now that you have broken the ice with the students in your class, ask the same questions of your lecturer, and break down the ice between students and staff.

UNIT 2 Learning with BTEC

In this unit, you will:

- find out about BTEC and BTEC qualifications;

- identify what is distinctive about BTEC Business and Finance programmes, and the reasons for taking them rather than other programmes.

What is BTEC?

BTEC is the **B**usiness and **T**echnology **E**ducation **C**ouncil, which is an organisation created by the Government to provide a system of work-related qualifications. These qualifications are nationally recognised by employers and educationalists throughout England, Wales and Northern Ireland. (Scotland has its own equivalent, called SCOTVEC.)

BTEC courses are available in hundreds of colleges and in some schools throughout the country. BTEC does not teach or examine the courses directly – this is done by lecturers. However, BTEC employs over a thousand moderators, who visit colleges to make sure that the courses meet the high standards set by BTEC.

Each year, BTEC programmes are chosen by around 180,000 students. They are available in a wide range of vocational areas: from nursery nursing to engineering; from catering to building; from fashion design to computing. The most popular, however, are in Business and Finance, with about 50,000 students a year. Of these 50,000 students, 30,000 are taking National level programmes. The BTEC National in Business and Finance, which is the programme that you have chosen, is, therefore, one of the most popular in Britain.

BTEC qualifications

BTEC qualifications stretch from GCSE-equivalent up to degree-equivalent qualifications, but unlike GCSE, A-level or most degrees, BTEC qualifications are 'vocational'. This means that they are designed to prepare you for particular kinds of careers, or to improve your job prospects, if you already have a job. These

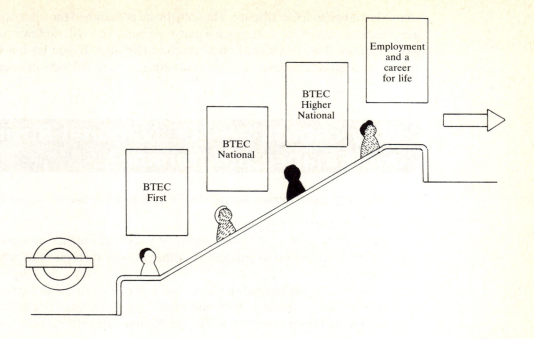

Figure 1 BTEC escalator of qualifications

qualifications are available at three main levels: **First, National** and **Higher National**. The three levels of BTEC qualifications are widely accepted as being equivalent to other well-known academic qualifications.

You can see from Figure 1 how BTEC programmes form a simple, easily understandable system of qualifications. You can enter and leave at a number of points, and when you have obtained a qualification at one level, this allows you to move on to the next level. If you get a job at the end of a qualification, or while you are taking one, you can move over from full-time to part-time study; if you leave a job, you can move over to full-time study. All of the time that you are on a BTEC Business and Finance programme, you are getting a broad-based preparation for a wide range of jobs, together with the chance to specialise in specific areas and slant your studies towards local job opportunities.

Part-time and full-time BTEC Business and Finance programmes

You can take a BTEC National in Business and Finance on either a full-time or part-time basis. A full-time course is known as a **Diploma** whereas a part-time course is known as a **Certificate**. The Certificate is designed for students who have jobs: because you attend college on a part-time basis you will do fewer subjects on the Certificate than you would on a full-time Diploma. If you have a Certificate and wish to get a Diploma, you can study for the extra subjects in another year.

Why choose a BTEC National Business and Finance programme?

There are three main attractions of a BTEC National in Business and Finance:

1 It can lead directly to a job.
2 If you already have a job, it will improve your performance at work and, therefore, your promotion prospects, either at your own place of work or elsewhere.
3 It provides a ladder of qualifications that you can climb, with each rung offering a route to further study and higher qualifications. You can climb from First to National and on to Higher National qualifications.

Activity 3   Why take a BTEC programme?

Read the following three case studies. Do they apply to you? Are these your reasons for taking a BTEC programme?

Jev: I'm leaving school next summer after my GCSEs and I want to go to college to do the BTEC National. I want to go into our family business after the National, but my dad wants me to join straight after GCSEs. My dad says you can't learn to run a business from books and if I'm determined to stay on I might as well do A-levels.

What do you think are the pros and cons of the three options that Jev has next summer?

1 BTEC National
2 A-levels
3 Straight into work.

Joanna: I'm just 18 and I've worked in an office for a firm of solicitors since I left school. I've got six good GCSEs. I like my job but they only really let me do the routine work. It's a large, busy office and I want to have more responsibility. I want a career – either with this firm or with someone else, if they won't give me the chance to get on.

What do you think Joanna would gain from doing a BTEC National in Business and Finance as a part-time student?

Alison: Next summer I finish my GCSEs and I should do pretty well. I'm not really sure what I want to do. I fancy a career in business – accounting maybe, as I'm good at maths. But I'm not sure and, anyway, I want to get a degree before I commit myself. So I'll probably do A-levels – I'll probably do Economics to get an idea about business. I enjoy academic study – reading, essays and so on.

Do you think Alison's choice is a wise one? Where could she get more information to help her to make up her mind?

Jev, Joanna and Alison are fairly typical of people who are ambitious and have to make a choice between working, studying full-time or studying part-time while in a job.

For Joanna, a part-time National will increase her knowledge and business skills so that she can take more responsibility in her present job or look for a job with more responsibility somewhere else. But her employer may not support her in doing the course, and she could have to fit in the course in the evenings after work.

Jev has a more difficult choice. He would learn a lot from two years on the National, which would stand him in good stead in the family business. But he has to balance this against losing income for two years and the fact that his father wants him to go straight into the business, as well as the experience that he could gain in two years at work. He could study for the course part-time but will have to convince his dad that it's worth doing, if he wants to get the day off to go to college.

Alison, like many school leavers, is unsure of what she wants to do. Even if she saw a careers counsellor, she might still be undecided. So, doing A-levels will probably make sense. It keeps her options open in case she doesn't want to work in business. Also, she seems to favour the academic style of A-levels rather than the more practical BTEC approach.

The three cases show that there is no correct answer to what choice to make between full-time work, full-time study and part-time study. It all depends on your personal circumstances and ambitions. The only certainty is that you should get as much information and advice as you can to help you to make up your mind.

BTEC programmes lead to jobs

Employers are very keen to offer jobs to people who show they are committed to work. Your BTEC qualification shows that you already know about business, are committed to working in a business and would settle into work quickly.

BTEC programmes involve employers

You will have help from employers while on a BTEC Business and Finance programme. BTEC programmes are designed by BTEC working with employers to ensure that what you are learning is up to date and relevant. Your tutors also work with people in local businesses to ensure that your BTEC programme is realistic and successful. Local employers help with many aspects – for example, they help in designing the programme, and are often involved in teaching and grading your work. Many employers provide work experience placements, and they allow you to visit their premises and interview their staff to find out what the job is really like. They also come into college to give talks and to help make sure that what you are learning really reflects the business world of today.

BTEC programmes are mainly taught by your lecturers, but employers are involved to ensure that the material is up to date and that you are well prepared to cope in the modern business world.

BTEC programmes are practical

You will cope better with working life and be a useful employee after having taken a BTEC Business and Finance programme because you learn the practical side of business as well as the theory. A BTEC programme is practical because of:

* the way you learn;
* what you do.

The way you learn

On previous courses that you may have taken, you may have learnt mainly from instructions given to you by a particular teacher, backed up by reading from textbooks and possibly researching for projects. Certainly, on a BTEC Business and Finance programme, you will learn from your lecturers, and you will have to do a lot of reading and researching. But the most distinctive feature of learning on a BTEC programme is that you will be 'learning through doing'. This means that much of your time will be spent tackling business problems and tasks, on your own or in groups with your lecturers to help and guide you. Your lessons will not be the kind where the lecturer stands up and talks to you for long periods of time while you take notes. Instead, you will be expected to take an active part, and many of your lessons will involve you carrying out business tasks or solving business problems. There will, of course, be times when your lecturers will give talks, but these will be kept to a minimum. The lecturer's main job is to prepare the business tasks that you are to work on, to guide you as you carry them out and to check that you have understood the main points. The lecturer will also grade your work when it is completed. You will find that, since you will have to participate in the lessons, you will be learning all of the time. It is a well-known fact that you learn more when you're actually involved, rather than being talked at (although it may be harder work).

Most students enjoy the freedom and challenge of this type of learning. They feel that it is more adult and find that there is a lot of variety in what they are doing.

What you do

The activities that you do on a BTEC programme will help you in your job. You will be required to do assignments, and may be required to run a business and take up work experience too.

Assignments

During a BTEC programme, you will complete what BTEC calls assignments. Assignments are pieces of work that you carry out at regular intervals throughout the course and on which you are graded as part of your assessment. These assignments are practical business tasks, partly to test what you have learned but also to help you to learn more about the topic. You, therefore, learn through doing assignments, and you will gain much benefit from working on and completing them to deadlines. Assignments take many forms: case studies, projects, role plays, reports, business games, simulations and many others that are designed to simulate the kind of tasks that you might have to perform in a real business.

Business simulations

The next best thing to working in a real business is to work in a practice or simulated business in college. As part of your BTEC programme, therefore, you will probably get the chance to learn how a business is run by working in a simulated business. In many colleges, students set up and run mini-businesses as part of their BTEC programme: secretarial students may work in the college training office; travel students may run a model travel agency; business and finance students may run retail businesses or a variety of services. The basic idea of simulation is to give you the maximum experience of dealing with realistic business tasks with the minimum of risk. You learn in college what it's like in the real business world and get the chance to do jobs that you wouldn't be given in a real business when you have just started work. It helps you to gain a broader understanding of what is involved in running a business, to develop practical skills, and to get a job and show an employer that you can take responsibility.

Work experience

If you are a full-time student, you will almost certainly gain some work experience on your BTEC programme. This will give you the chance to put the skills and knowledge that you have gained at college to work in a real business, and to acquire other knowledge and skills from the placement. You may also have to arrange visits to employers' premises to gather information for projects and assignments.

Practical activities

The BTEC Business and Finance programme will introduce you to up-to-date business practice, and you will be expected to do many of the practical things that you would have to do in a real business. For example, you will be required to do tasks such as:

- using office equipment;
- using computers;
- writing business correspondence;
- organising and running meetings;
- preparing promotional material;
- preparing accounts;
- keeping financial records;
- preparing job adverts and job descriptions;
- dealing with legal problems.

Activity 4 gives you an example of the kind of activities you might undertake on your BTEC programme, especially at the start in your induction programme.

Activity 4   Handbook of college services

Every college provides a wide range of services for your educational, leisure and personal needs. If you are to get the most out of your time at college, you need to know what services are available, and where and how you can use them.

Working with others in your group, design and produce a handbook of college services for the students on your course. You will need to visit all of the services to find out when they are available and if there are any restrictions on your use of them. Your handbook should include a map of your college with the services marked on it.

To get you started, here are some services that can be found in most colleges:

- Student counselling service
- Student common room
- Student union
- Sports area
- Refectory

- Teaching areas – base room and main classrooms that you use, computer facilities, library/learning resources area, keyboarding rooms, Business Studies resource area, business simulation areas (shops, training office, travel agency and so on), audio visual aids service

- Staff areas – Principal, Head of Department/Section, course tutor/co-ordinator, department office and secretary

- Safety – fire exits and assembly areas, fire extinguishers and alarm points, first aiders and first-aid boxes.

BTEC programmes are designed to suit employment locally

Most people get their first job in the area in which they live and study. BTEC Business and Finance programmes are, therefore, designed to prepare you for the local employment scene as well as the national scene. This is made possible because BTEC issues flexible 'guidelines' to centres on what to teach, unlike GCSEs and A-levels where there is a national syllabus to be taught and national exams are set. The centres can then use the BTEC guidelines to design their programmes with local employers and agencies so that they are relevant to business in the local area, as well as in the country as a whole. This means that a National in Business and Finance in a popular tourist area like Blackpool or Brighton will have a different slant from one offered in a traditional industrial area like Chesterfield or Barrow. Students on both of the programmes will cover the same content, but the work in classes, the projects and the assignments they do, will be designed to help them to get jobs in the local industries, as well as in the country as a whole. This is why BTEC doesn't set examinations that everyone has to take. Instead, BTEC centres are free to set work that meets the requirements of the BTEC programme and the needs of the local situation. Don't think that this makes a BTEC programme easier than A-levels though, because your work is supervised by a BTEC Moderator who visits the centre to inspect students' work and to make sure that the programme comes up to the national standards that BTEC has set.

BTEC programmes develop business skills and qualities

A BTEC Business and Finance programme aims to prepare you for what it's really like in business. You will be expected to do many of the things that you would have to do in a business as a normal and regular part of your BTEC programme. You will be tackling real or realistic business tasks or problems both in and out of class. You will learn about the theory and practice of business, but in addition to this you will be developing certain types of skills, which will help you to be effective at work and to adapt from one situation to another. Examples of these skills are:

- working with other people;
- dealing with information;
- presenting information;
- making decisions;
- solving problems;
- communicating effectively;
- organising.

BTEC programmes improve your performance at work

If you have a job and are a part-time student, doing a BTEC Business and Finance programme will help you to improve your skills and qualities, which will help you to improve your performance at work and, therefore, help your promotion prospects. You will be able to practise these skills at work straight away. You will also have the opportunity to do projects or assignments that are related to your job or in areas of your business that interest you. This will help you to gain a wider understanding of the business in which you work and will be beneficial to your employer.

What can you do after your BTEC programme?

Full-time education

Many BTEC students continue in further full-time study to get higher qualifications. If you choose to join or continue in full-time education, you can proceed to the next level of BTEC qualification. From the National, you can progress to the Higher National Diploma, which is a two-year course. Or, if you get very good grades on the National – bare passes won't be enough – you could take a degree at a university or polytechnic. Your broad-based National would equip you for a Business Studies or Economics degree, and you could also look at more specialised degrees in Law, Accounting, Marketing, European Business or Retail Management.

Employment and further qualifications

Many BTEC students want to go straight into a job at the end of their BTEC programmes. They find that BTEC qualifications are widely accepted by both large and small employers. Manufacturing companies, banks and building societies, retail chains, as well as central and local government, all recruit large numbers of young people straight from BTEC programmes.

Once you are in a job, you may want to advance your career by getting further qualifications by studying part-time. You may choose to continue with BTEC by taking a two-year Higher National Certificate or some short, sharp highly specialist course leading to the award of BTEC Continuing Education units, or you may want to study part-time for a degree. If you want to specialise in the area that you're working in, you may want to get the professional qualifications for that job, such as accounting or marketing, or as a Chartered Secretary. Over 150 professional bodies will accept a BTEC qualification as an entry qualification to their professional status. Often, when you have already studied a subject with BTEC, you will be given credit for this, and you won't have to study it again for the professional qualification. So, although your BTEC qualification is broad based, it can serve as a step towards more specialist professional qualifications.

Self-employment

Today, more and more people are choosing to work for themselves and run their own businesses. Some go it alone as sole traders, whereas others join together in a partnership or co-operative. You will find, if this is the route that you have decided to take, that you will have acquired and developed on your BTEC programme a lot of the knowledge and qualities that you will need to make a success of your own business. You should have the confidence in your own ability to cope with the realities of business life, as well as a good knowledge of all of the sources of help and information that are needed to make a success of a small business.

Therefore, the time that you spend on a BTEC National Business and Finance programme is time invested. You invest because BTEC is a proven route to employment immediately on completion, or after having taken further qualifications. Your BTEC qualification has been designed to equip you for both employment and further study, so you won't be short of choices.

Activity 5   Working for yourself or for others?

One of the fastest growing areas of employment is self-employment. Many people who have lost their jobs or have become bored with them decide to set up their own businesses. Often, it's a chance to do something that they have always wanted to do but didn't have the time for. That said, it's a risk: many businesses go bust in the first year of operating.

Think about the attractions of self-employment and compare them with working for someone else. You may know some people who work for themselves; ask them what they think. Using Table 3, tick the category (self-employment or employment) that best matches each statement.

TABLE 3   *Comparing self-employment with working for someone else*

Statement	Self-employment	Employment
Easier to get started working		
Better pay		
More security		
More challenging		
More satisfying		
More variety		
More pressure and stress		
More freedom		
Better holidays		
More people to work with		
More routine		

Which option appeals to you most? Do you think that you will feel the same in five years? Discussing this with others in your group will be very useful. You'll see that people have different ideas about what working for themselves means and that some people may be better suited to it than others.

UNIT 3 Learning business and finance with BTEC

In this unit, you will:

- discover what you will learn on a BTEC National;

- find out about the distinctive BTEC way of learning, called 'learning through doing';

- find out how to review your own performance and improve in the future;

- find out about the ways you learn and the different people who will help you to learn on the National.

BTEC modules – built around business

Many of the exams that you have taken previously will have been based on long established and recognised subjects – English, Maths, Chemistry and so on. To get your qualification, whether at GCSE or A-level, you study a selection of topics from within one subject. Your BTEC subjects, called **modules**, are quite different. Each module is built around an area of business activity – 'Business Environment', 'Human Resources', 'Financial Resources'. The issues dealt with in these areas of business don't fit neatly into subjects like Maths, English and so on. Understanding 'Business Environment', for instance, means that you need to study parts of lots of subjects. You study some Economics, Law, Politics and Sociology, and will also do some Maths and probably Computing. So, the various modules on your BTEC programme will have names that are new to you. Some of these modules are compulsory, while for others you have a selection from which you can choose. The compulsory modules are called the **core**; those where you have a choice are called **options**. In addition to the core and option modules you also have to study and develop your common skills.

Core modules

There are eight compulsory modules which make up the core and together they are called 'Working in Organisations'. What you learn in these modules is of great value to you whatever career path you take in business, that's why they are compulsory.

Business Structures and Goals

In this module, you will learn about:

- the different types of business organisation and their different goals;
- the different types of goods and services provided by public and private sector organisations;
- how organisations are structured to carry out their operations.

Activity 6   Summarising an organisation's structure and goals

Using an organisation with which you are familiar, such as your school/college, or your own place of work, produce a simple summary of its goals, the goods or services it provides, how it is organised and who owns it. You will find it useful to compare your summary with that of someone else who has looked at a different organisation.

Business Environment

In this module, you will learn about:

- the main industrial sectors in the UK;
- the role of government in economic affairs;
- how the financial system works;
- patterns of trade and investment between the UK and the rest of the world including the importance of the European Community.

Activity 7   'Swotting' an organisation

A 'SWOT analysis' identifies the strengths and weaknesses of an organisation and the opportunities and threats facing it. This type of analysis is useful because it encourages people to think about the organisation in its environment. Also, because the environment is always changing, SWOT helps you to identify these changes, and enables you to adapt to new opportunities and threats on the basis that 'as one door closes another door opens'.

Carry out a SWOT analysis using Table 4 to identify the strengths, weaknesses, opportunities and threats of an organisation of your choice. If you are a full-time student, you could choose your college. If you are a part-time student, you may wish to use your place of work. Alternatively, you could select any organisation that interests you. To get you started, some suggestions have been included in the table, taking the college as an example.

TABLE 4   *SWOT analysis*

Strengths	Weaknesses	Opportunities	Threats
Wide range of courses	Location of college – not on a main bus route	More adults need training	Drop in numbers of 16 to 19 year olds

Marketing Process

In this module, you will learn about:

- how the market system based on supply and demand operates in theory and in practice;
- how businesses go about marketing their goods and services to customers/ clients.

Activity 8   From 'swotting' to 'marketing'

Using the same organisation that you 'swotted' in Activity 7, now identify its major customers/clients and the methods used to advertise and promote its goods/ services.

Physical Resources, Financial Resources, Human Resources

These three modules need to be considered together as they deal with the essential resources needed by any business:

- physical resources such as land, buildings and equipment;
- financial resources – that is, money or credit;
- human resources, which means the employees who make the business work. In any organisation, these people are the most important resource.

Every business needs all three types of resource but each business will need them in different combinations. For example, some businesses are capital intensive and use a lot of equipment while others are labour intensive using many employees. You will study how to make the best use of resources, whether money or people, as the next two activities show.

Activity 9   Your personal budget

Draw up a balance sheet of your own income and expenditure over the last week using Table 5. Make a note of the main types of income and expenditure and compare them with those of another member of your group. What lessons can you learn from this?

TABLE 5   *Balance sheet of income and expenditure*

Income	Expenditure
Total	Total

Activity 10   The importance of people

People are an organisation's most important asset. People can give good or bad impressions to customers and clients of the organisations in which they work. To demonstrate how people in organisations can create good or bad relationships with customers and clients, read the following conversations and explain how you think each person felt immediately after the conversation.

Customer complaint 1

Customer: Excuse me, I bought these shoes last week and look at them, the sole is coming away.
Sales staff: Have you got your receipt?
Customer: No, I'm sorry I haven't, I really didn't expect to have to keep it.
Sales staff: I can't do anything without your receipt.
Customer: But they are faulty.
Sales staff: Well, if you haven't got your receipt, I don't know where or when you bought them. I mean you could have bought them months ago, they do look well worn.

Customer: That's exactly ...

Sales staff: (interrupting) ... and, without your receipt, how do I know you bought them from this shop?

Customer: They have your name on the ...

Sales staff: (interrupting) ... that's why we give you a receipt, you know. You could have bought them from anywhere, Dolcis, British Home Stores, Marks & Spencers, Lewis's, Curtess or anywhere.

Customer: I want to see the manager. I know the law.

Sales staff: You're talking to the manageress and the law says you need your receipt. I can't do anything without it – it's proof of purchase.

Customer: I'm going to the Trading Standards.

Customer complaint 2

This is a telephone conversation between a customer and a mail-order firm.

Customer: Hello, my name is Mrs Patel. I am owed £24 and have waited months but have still not received my cheque.

Staff: Can I have your account number please?

Customer: Yes, it's 567012 S.

Staff: I'm sorry, I have nothing on record.

Customer: I returned some goods to you and I am now owed £24. I phoned a few weeks ago and I was promised it would be sent, but I have received nothing.

Staff: I'm very sorry you have had this problem. Can you tell me what goods you returned to us and when please?

Customer: It was a coffee maker and it was returned in July.

Staff: Thank you. Are you able to give me your telephone number and I will look into it and ring you back this afternoon? I really am sorry you have had this problem. I will look into it immediately.

Administrative Systems

This module deals with some of the more basic practical aspects of business – the paperwork, procedures and technology used in the office. You will learn about:

* who does what within the structure of business organisations;
* methods of communication at work;
* how to use computers for administrative tasks;
* the main types of business document used.

Activity 11 Organisation chart

Draw up an organisation chart for a business organisation with which you are familiar, indicating key personnel and areas/divisions of work.

Innovation and Change

The only certain thing about business is that new products/services or new ways of operation are essential for survival. So, in this module, you will learn about:

- the main factors forcing business to change;
- how innovation and change, such as computerisation, affect work and the workforce;
- what businesses will look like in five to ten years' time.

Activity 12 Change

If you are a part-time student, find out how the organisation in which you work has changed in the last 30 years, and explain why it has changed.

If you are a full-time student, find out how your local shopping centre has changed in the last 30 years, and explain why it has changed.

Common skills

In Unit 2, you learned that BTEC programmes develop your skills as well as your knowledge. You may have come across skills assessment before, on a GCSE or DOVE/CPVE course, on an internal training course at work, or on a course outside work such as a first-aid or sports coaching course. The basic idea of developing skills on this course is quite simple:

- There are some general skills or abilities needed for any job in business: BTEC identifies seven such general skills.
- These skills are a part of everything you do on your BTEC programme and should not be just tagged on or treated separately. Therefore, although common skills can be called a module, you should not be taught a separate subject called common skills. Even if you are, you will also be learning about and using these skills when you are doing the other modules. For example, if you are learning about how firms select staff, you will also be learning about communication skills.

The seven skills identified on the National are:

1 Managing and developing yourself – organising yourself to carry out tasks and get the best out of your time.
2 Working with and relating to others – dealing with customers/clients or working with colleagues, often as part of a team, to tackle problems.

3 Communicating – written, verbal and non-verbal communication to different types of people.
4 Managing tasks and solving problems – analysing and solving business problems on your own and within a team.
5 Applying numeracy – handling, understanding and presenting figures.
6 Applying technology – using computers and other new technology to carry out business tasks.
7 Applying design and creativity – using your imagination to create or improve something, such as a new document or the layout of work area.

These seven skills are in no particular order – the first on the list is no more important than the last.

Activity 13 A manageress at work

Read the following extract from the diary of Annetta who is a marketing executive with a large multi-national company and identify which of the common skills she is using.

Monday: AM Meeting with the sales manager to discuss the monthly sales figures.
          PM Business lunch with Mrs Taylor to discuss the contract with her firm.
               Appointment with Mr Smythe (Cardon Ltd.) to discuss his complaint about quality control.
Tuesday: AM 9.30 Disciplinary hearing – Mr Rhodes (lateness).
               10.30–12.00 Read report on Saudi contract from Peter Jackson and draft digest of main points for the Board Meeting.
               Fax expenses to Head Office.
          PM Shortlisting for office manager post.
          Eve 'Exporting and You' seminar at the Chamber of Commerce.
               See JF about marketing seminar.
Wednesday: AM Meeting with general office staff to discuss layout of offices in new building.
          PM Business to Business Exhibition – 12.20 train.

There are many different ways in which you will develop these skills. When you start your course, you will be given the chance to show how good your skills are, and to identify with a tutor where you need to improve and what you are good at already. During the rest of the course, you will be taught about the skills, will practise them and will be assessed on them in your assignments.

Option modules

You don't have any choice about doing all of the core modules and common skills – they are so fundamental you can't miss any out. With your option modules, you do have a choice from among a number of options offered in your BTEC centre.

The options let you look in more depth at specific areas of business where you might want to work. In Financial Planning and Control, International Marketing and Business Information Technology, the options pick out aspects you have already met in the core and take them to greater depth. You may have a particular career already planned in banking, personnel or finance, or you may want to try different areas. The options allow you to do this. You can put together a package of options to suit your career interests.

If you are part-time and doing a Certificate, you take four options; if you are a full-time student taking a Diploma, you take eight options. There is a wide range of options, but not all BTEC centres can offer all of them. Some of the more common ones are discussed here to give you an idea of what they are about.

Business Information Technology

This module is useful if you want to know how to make computers work in business, and includes word processing, spreadsheets, data bases and desk top publishing. You will use the types of programs that you would come across in business.

Business Law

Business Law deals with the way a business is set up, and the legal aspects concerned with how the organisation trades with other organisations and deals with its customers and staff. You not only learn about the law but also the practicalities of what to do if things go wrong.

International Marketing

Marketing is often described as 'getting the right product to the right place at the right price in the right quantity at the right time'. You will learn how to conduct market research in order to find out the needs of customers. You will also learn about how to price, promote, advertise and sell products in international markets.

Personnel Policies and Procedures

This is a useful option for you to take if you wish to work in the personnel department of a large firm or in a small business, or if you want to start your own business and will be involved in employing people. In this option, you learn about: keeping personnel records; legal aspects of employing and dismissing people; motivating and training staff; interviewing skills and dealing with staffing problems; and the increasingly important area of equal opportunities in employment.

Languages

Languages are becoming increasingly important in business life, especially with the creation of the European Market. Language units concentrate on how you might use the language at work – in writing, telephoning or meeting overseas customers or suppliers, arranging business travel and knowing how to cope in another country while on business trips. You can study languages such as Italian, Spanish or German, but French tends to be the most popular.

Activity 14   Finding out what options are available

Find out which option modules are offered by your BTEC centre and are not listed here. Find out what you will study on each of those modules and what kind of jobs they might prepare you for. Using the following table, make a list of your findings.

Option	Content	Jobs prepared for

Choosing your options

When you have found out about all of the options that your BTEC centre is offering and what they are about, you need to decide which options you are going to take. However, it is sometimes difficult to decide which options to choose. Here are some tips to help you choose:

- Find out what kinds of job the particular modules might lead to and choose the modules to suit the type of career you would like – you started this in Activity 14.
- If you cover a topic that particularly interests you in the core in the first year, you may be able to choose options in the second year to learn more about this area.
- If you are a part-time student, you should discuss which options are most useful to you with your employer. Your employer may have very definite views on which are best for you.
- You may do something on your work experience that you would like to follow up in an option.
- Seek advice from your tutors and careers officers.

If you still can't decide, then try the following activity.

Activity 15 Which options do I choose?

If you are still unsure about which options to choose, make arrangements to visit some people who work in jobs that you think would interest you. Ask them about the job, what it involves and what you need to be good at to do the job well. This may help you to decide which options are best for the kind of career that you would like to have.

Learning on an integrated programme

Your BTEC National is a programme of closely related modules, not just a set of unrelated subjects. The main ingredients of the core – the core module and common skill areas – are carefully designed to fit together to give you a complete picture of business. They fit together to form a whole. The options spread out from this core, allowing you to specialise or take further the areas introduced in the core. In the core, you might have been interested in the section concerning customers, and so you may decide to take it further by taking the International Marketing and Business Law options.

You will also find that you will have to use information and skills that you learned in one module to help you with another. This isn't an accident. It is one of the distinctive features of BTEC programmes that makes them quite different from A-levels or even GCSEs, where each subject is quite separate. In business, it is very rare to find a job that only needs you to be good at one thing, especially if you are in a supervisory position. You usually need to draw on a whole range of knowledge and skills to get a job done to a high standard. At work, you do not deal with problems that are so narrow that they involve only one specific subject. For example, a redundancy problem at work will involve knowing about financial and legal aspects concerned with the redundancy, dealing with unions, negotiating, writing letters to the employees who are to be made redundant, calculating redundancy payments and so on.

Your BTEC programme – finding your way through the maze

Although your BTEC programme seems set out for you in the modules and in the various activities and assignments set by your tutors, you will have your own programme, or pathway through the course. No two programmes or pathways will be exactly the same. For example, from among the options, some students may choose Business Information Technology, while others may choose Business Law. If you are at work, you may not have to do an activity or part of an assignment in college if it is something you already do at work. Because everyone has different strengths and weaknesses, each person will have to put in that extra effort in different areas or include more of something in their programme – perhaps more work on maths or learning how to use a computer.

So, an important feature of your BTEC National is that you have to agree your own programme with your tutor and review it regularly to ensure that you are on track. Often this is described as 'negotiation' with your tutor and in some colleges you may have to sign a 'learning contract' setting out that you have agreed the programme and promise to work hard to complete it successfully. This is not a legally binding contract but it is a useful exercise as it forces you to decide your programme and commit yourself. It also gives you a flavour of the real world of work where employees have contracts setting out their responsibilities and rights, which they accept by signing the contract.

Although each student's pathway will be slightly different, nevertheless there are some clear stages each student goes through. These are:

- Joining the programme: when you start the programme, you will be assessed to decide what you already know or what relevant experience you already have.
- Deciding your programme: you will pick out your own pathway for each stage of the programme and write down the first of your 'action plans'.
- Learning on your programme: you will complete lots of different activities and assignments from which you learn and are assessed.

- Reviewing: you will review how you have performed on each assignment and update your action plan for the next steps along your pathway.

Joining the programme

When you join a BTEC programme, your tutor will need to identify what knowledge, skills and experience you bring with you. You may have examination passes or have trained in areas highly relevant to the course, for example in keyboarding, IT or Health and Safety. Or you may be at work using a computer, or in a Marketing Department, or handling invoices and other financial documents regularly. You may have experience which is nothing to do with business but which has developed useful knowledge and skills. For example, organising a local football team or helping to run a Girl Guide pack may develop useful skills in 'communicating' or 'working with and relating to others'.

Every student brings something useful to their programme and BTEC requires that tutors find out what this is and give credit for it. This is called 'accreditation of prior learning' (APL) and should take place right at the start of your programme. If you have done well at something in the past and can provide evidence of this, you could be allowed to miss certain parts of the programme. The idea is that you have already covered some elements of your programme and should not have to repeat them. For example, if you have recently been on a Health and Safety course at work, you could gain credit or APL for this and not have to do it on your programme. Or if you work in a Marketing Department, you may be given credit for some parts of 'Marketing Process'. In order to be given APL, however, your prior learning or experience must be fairly recent and must be up to the level of the BTEC National. Equally important, you have to provide evidence to back up your claim for APL – your tutor won't just accept your word that you have some prior learning or experience.

Activity 16 Accrediting prior learning

Cajetan, a 30-year-old office worker, faces redundancy because of the impending closure of his firm. He has lots of experience of working in a business but few qualifications. To improve his job prospects, he wants to get the National. With his experience, listed in the following table, for what areas of the programme might he be able to claim APL and what sort of evidence would he need to provide? To get you started, an example has been given.

Experience	Evidence of competence required
1. Produced word-processed reports	Sample report with note from employer that it is his work
2. Dealt with customer queries by phone	
3. Supervised three office staff	
4. Responsible for health and safety in office	
5. Union representative for three years	
6. Secretary to Church Parish Council	

Deciding your programme and action plan

Once you have worked out your starting point and gone through APL, you can select your pathway through the core, options and common skills. You will also draw up an action plan based on your personal strengths and weaknesses. This is your own personal plan of what you are going to work on over the first period of the course, together with targets and deadlines for you to hit, and a date when you will review your progress and update your action plan. All this will be done with your tutor: it may be a new experience for you to discuss your targets and progress with someone in such detail, although if you have experience of TVEI and profiling you will be quite familiar with this approach.

The things you put in your action plan may be quite personal, and not things you would want people other than your tutors to know about. It isn't easy to write an action plan, but by putting plans on paper, you are commiting yourself to achieve something that is useful to you. In addition, the thought of having to meet your tutor on an agreed date in the future to review your progress is a useful discipline for you. Ultimately, though, it is your action plan, not your tutors, and you are responsible for updating it.

To summarise, on your National programme, in order to get the full BTEC Diploma, you must take and pass (or gain credit through APL):

* the core: 'Working in Organisations'
* common skills
* options.

So, the overall structure of the programme of a Certificate or Diploma student will be as shown in Figure 2.

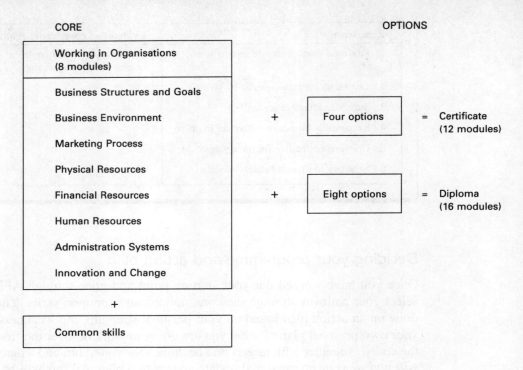

Figure 2   Overall structure of the programme

How do you learn on the National?

In Unit 2, you saw that employers look for well-educated staff who are competent to do a job. They are looking for people who not only know about business but who are competent to work – not just 'thinkers' but 'doers'. So, your BTEC programme is taught in ways that develop your ability to 'do' as well as to 'think'. This BTEC approach to learning is often described as 'learning through doing'. Examples of the things that you will do on the National are practical assignments, group work, work experience, practical work, computing and research. It isn't an easy way to learn, and at times you may come to see the National as a bit of an obstacle course, with one assignment or group task after another. You might even wish there were fewer things to 'do' and more emphasis on note taking or just following simple instructions from your tutors. But it is an interesting way to learn, and you can learn to do more this way than you would do just by note taking. It is important, therefore, that you fully understand the concept of learning through doing, so you know why you work in this way and can go about it successfully.

A good way to understand learning through doing is to contrast it with the more traditional ways of learning which everyone has experienced at some time or another. Although every tutor may have a distinctive teaching style, some courses, such as most A-levels, degrees or professional qualifications, push the teaching staff towards traditional teaching methods, such as lecturing. Other courses such as BTEC, and GCSE to some extent, push teaching staff towards getting students to learn through doing.

Activity 17 What is expected of a BTEC student?

You can see from Table 6 what is expected of students and tutors on a course using traditional teaching and learning methods. Consider what is expected on an A-level course and, if possible, discuss this with fellow students and your tutor. Then complete the table, listing what is expected of you and your tutors on the National.

TABLE 6   *Comparing traditional teaching/learning methods with the BTEC approach*

Traditional course	National course
Expected of the tutor: 1 Knowledge of the subject 2 Knowledge of the exam requirements 3 Ability to pass on a lot of information	Expected of your tutor: 1 2 3
Expected of the student: 1 Interest in the subject 2 Note-taking skills 3 Extensive reading 4 Good at exams 5 Ability to work independently	Expected of you: 1 2 3 4 5

When you have worked through this activity, you will realise that a lot more is expected of you and your tutor on a National course than on a traditional course. The A-level tutor and student have to work hard but are relatively passive. The tutor follows someone else's syllabus, giving information and ideas, and preparing students for nationally set examinations. The student is largely the receiver of information from the tutor or extensive reading, which must be understood and reproduced in exams. In contrast, the BTEC tutor shapes the content of the course, designs the assignments and assesses the students subject to checks by the BTEC

Moderator. The contrast between an A-level student and a BTEC student is equally sharp. The BTEC National student has to tackle lots of varied practical assignments, often working as a member of a group, and is assessed continuously, not only on the content of the assignment but on his/her skills performance, general enthusiasm and initiative shown in tackling assignments. So, the National is very much what is called a 'student-centred' course; it is centred on you, the student, learning through doing. Your tutors set work, give guidance while you do the work, mark your work, and give you feedback on how you are doing and how you can improve.

Activity 18 Benefits of learning through doing

Central to the BTEC style is the belief that learning through doing is the best way to prepare for a business career, although it is not without its pitfalls if done badly.

List the benefits and pitfalls of learning through doing using Table 7, based on your experience of investigating changes (see Activity 12) in either your place of work (if you are a part-time student) or the local shopping centre (if you are a full-time student). A few points have been included to start you off.

TABLE 7   *The benefits and pitfalls of learning through doing*

Benefits	Pitfalls
1 More interesting work	1 Time consuming
2 More variety	2 Lots of work
3 Develop business skills	3 Group work can be hard
4	4
5	5
6	6

Your own experience will have told you the benefits and pitfalls of learning through doing. Some of the points you may have thought of are as follows. Learning through doing helps you to increase your confidence because you can do the various tasks assigned to you. Also, because the tasks are realistic business problems, your confidence to do a real job increases too. When you 'do' rather than listen or watch someone else doing a task, you understand more and also remember more. Doing increases your skills so that your job prospects increase, and you fit better into work. Learning through doing is also more interesting than just sitting and listening. Some of the pitfalls, however, are that you need to make an effort and show initiative in working on problems. Learning through doing is time consuming, especially if you are working in a group; you will find that you will be spending a lot of time doing assignments, although it is time well spent. Working on assignments can be frustrating, because you may need to attempt a problem time and time again until you succeed. Because your tutor will discuss your work with you, you need to be able to accept criticisms of your work, which can be difficult if you have put a lot of effort into it.

How do you learn through doing?

It is possible to unravel learning through doing a bit more systematically to see the various stages of your learning and what is happening at each stage. Table 8 shows the five main stages of learning that you go through, taking as an example an assignment involving a role play, by part-time Certificate students, of management and union negotiations over a wage claim.

TABLE 8   *Stages of learning through doing*

Stages	Example
1 Prior experience – you will already have knowledge or experience to bring to the assignment.	1 Prior experience – you may be in a union and have some experience of how wages are negotiated, or you may have picked up a lot about unions from the media.
2 Tutor sets practical problem and briefs students.	2 Tutor distributes information about the company, decides who will play which part, and spells out the rules of the game and the time available.
3 Students carry out research and tackle a business problem.	3 Students read up on wage bargaining to prepare for their own roles and act out the negotiating process of the wage claim.
4 Tutor assesses performance and debriefs students.	4 Students get their grades but also discuss with the tutor and among themselves how they performed, why things went wrong or right, and what they need to improve on in the future.
5 Tutor and students generalise – they pick out the points of lessons from the assignment, which may apply generally or be transferred or used in other assignments. This then becomes part of the prior experience for later work.	5 Tutor suggests further reading to students and leads discussion to draw out general points – for example, on negotiating, communicating or handling conflict – which students will encounter again and will be able to transfer to different situations or assignments.

You may have thought that learning through doing only really involved stages 2 and 3, since these seem to be the main 'doing' stages, but that would be a dangerous view. Doing something is not sufficient to ensure that you learn anything from it. Some people never learn but keep on making the same mistakes over and over again. For you to learn from doing anything, you need to go through the other stages, especially stages 4 and 5. After an assignment has been completed, you need to reflect on the lessons learned, usually with someone else to prompt you. Unless stages 4 and 5 are worked through, you may not realise that the specific point you learned in the assignment is an important general principle or idea that will be useful to you later – and you may, therefore, forget it quickly.

Reviewing what you have done

'Well, that's it – another assignment out of the way. Now, what have I got to do next?' You'll probably say something like this quite often during the course – and there'll always be something else to do when you've finished one piece of work. But you know that handing an assignment in isn't the end of it. To get the full value out of an assignment or anything you have done, you've got to review that piece of work.

Tips for reviewing and generalising

Reviewing with your tutor

In the case of an assignment, you will receive a grade from the tutor together with some feedback on what you've done: the tasks you did really well, the tasks you could improve and some tips on how to do even better next time. This is useful as it should mean that you know what you have to work on in future assignments to get higher grades. If you don't understand or agree with what the tutor has said, or you want some more detail, don't hesitate to ask. Tutors will be happy to tell you why they've given you the grade that they have and to explain where you could improve.

With other pieces of work, you should also feel free to ask the tutors what they thought of the work that you have done. They'll be happy to discuss it with you and also glad to see that you have taken the time to discuss your work with them.

Review the work yourself

Try to consider what you did and what you got out of it. There are three questions to ask yourself:

1 What happened?
2 So what?
3 Now what?

What happened?

What did you actually do in the piece of work? For each piece of work, you will have to use some knowledge and some of the skills that you are learning on the course. What you need to do is to consider what knowledge and which skills you've used.

So what?

Once you've identified what was actually involved in the piece of work, there are a couple of extra questions:

• How well did I do these things?
• Were there any other things I could/should also have done?

So, for each of the things that you have on your list, you need to decide how well you did – the feedback on your work will help you with this.

Now what?

What you need to do now is to work on your action plan, listing the things that need a bit of work, the things that need a lot of work and the things that you haven't done at all. This is sometimes called **target setting**. So, if you know, for example, that you didn't use a computer as effectively as you could have done in work that you have handed in and had graded, you may put this in your action plan as a target. Meeting the target may mean that you have to ask the tutors how to use the program more effectively or that you need to put some time aside to practise.

An action plan will often take the form of a list, as shown in Table 9. Some suggestions have been added as examples. As you find more things that you need to work on, you can add them to the list, and you can cross off the ones that you have achieved. Some things may stay on the list (such as 'learn to spell') while others will be crossed off fairly quickly.

TABLE 9   *An action plan*

Target	**How to reach it**
Use the graph-drawing facility on Supercalc 4	Practise on the machine at work Ask Alan to show me Use it for next assignment

Ways you learn

As mentioned earlier, there are various ways in which you learn through doing. This section deals with these in more depth. You learn through:

- assignments
- business simulations
- work
- working in groups
- the induction to the course.

Learning through assignments

By now, you will have realised that assignments are at the heart of your course. Assignments are not tests set by your tutor to see if you have learned something after the topic has been covered in class. They are business problems that you learn from as you tackle them. As you tackle them, you develop your business skills as well as your business knowledge. Most assignments are in the form of a business problem, and you will be expected to work on the problem as if you were someone involved in the business – a manager, office supervisor, union official, customer, finance or computer expert and so on. You tackle the problem from the point of view of the position you are given, and you will be assessed on that basis. Assignments will usually involve some written work, but you may also have to give a talk, chair a meeting, work with others, or use business equipment such as a computer or a telephone, or several of these in one assignment.

Learning through business simulations

A simulation is an artificial situation that is designed to give a 'feel' of a real situation. Aircraft pilots train on flight simulators, police train in simulated riots or traffic accidents, and there are many computer programs available for use as business simulations. The key point about a simulation is that it is realistic, although it usually simplifies the real situation, and so allows you to try out things safely in a variety of circumstances and answer 'what ... if' questions. You can find out what would happen if you tried something by testing it in the simulation. On your course, you may be involved in simulations quite regularly, and many of your assignments may take the form of simulations. Computer software is available to let you simulate decisions, especially in areas of finance and marketing on, for example, what prices a firm should charge or how a firm should react to a sudden fall in the value of the pound.

The most extensive form of simulation, now very common on National courses, is the mini-business, where students set up and run a small business in college and then close it down at the end of the year. Detailed advice on how to run a mini-business simulation is provided in Unit 7.

Learning at work

If you are taking a Diploma, you will spend some time on unpaid work experience. You will probably be familiar with work experience from school or from having had a part-time job. Work experience is very important, and your performance is usually assessed. You benefit from it because:

* It ensures that you have recent and relevant experience of working in business. It may be very different from the image that you have of it. Three weeks working in a bank might convince you that this is where you want to work – or it might convince you of the opposite!
* It gives you the opportunity to put what you've learned into practice, or at least see others put it into practice. You'll see the skills and the knowledge that you've gained being put to work every day.
* It helps you to gain information on the way that a particular business works. This will be valuable later in the course. In addition, if you want to know what another business is like, then possibly someone else has been placed in that sort of business and can tell you about it.
* It allows you to experience the atmosphere of the workplace and what is expected by way of attendance, dress, behaviour and so on. It's not like being at college!
* It may help you to get a job – maybe even with the firm that you are placed with.

Your work experience may be organised in a number of different ways, and each way has a number of pros and cons. You may go on:

* Block release – where you go out to a business for a block of time, normally two to three weeks, and get really immersed in the work.
* Day release to work – where you go to a business one day a week over a full college year and see how things develop in a business over the year.
* Day release plus block – where you normally go to a business for a block of time, perhaps a week near the start of the course, so that you get used to the organisation, and go there one day a week. In this way, you get the best of both block and day release arrangements.

Whatever system your centre uses, you will have assignment work to do, based on your work experience placement. While on placement, you will be visited by your tutors to see how the work placement is working and to liaise with the employer. In Unit 7, you will find more detailed examples of assignments associated with work experience.

Many students find work experience to be one of the most valuable experiences of the course. On the other hand, many employers regard work placements as an opportunity to seek out potential employees.

Learning in groups

Everyone at some time or another has been a member of a group, working to achieve some target: raise money for charity; organise a school disco; arrange a holiday with friends. Many people are not prepared, however, for the extent of group work on the National. Even people who have experienced learning through doing in the past are surprised at how much 'learning' and 'doing' is involved on the National as a member of a group. It shouldn't surprise you, however, that a course that deals in realistic business assignments expects you to work with other people. Most problems at work are tackled by several people working together, and even the day-to-day running of a business requires people to co-operate with each other and work as a team. Not surprisingly, the skills of 'working with others' and 'communicating' are identified as two of the seven key common skills for your course.

Activity 19 The pros and cons of group work

From your own experience of working in groups, select a group activity that was successful and one that went badly. Consider why one group activity was successful while the other was not, and then list the advantages of group work, as well as the problems, using Table 10.

TABLE 10   *The advantages and disadvantages of group work*

Advantages	Problems
1 Share the workload	1 Some people slack
2	2
3	3
4	4
5	5

The following situations illustrate further the problems that can occur when you work in groups:

- You are working in a group with three other people. Two of them you like and get on well with. In fact, you are good friends outside college and you go out a lot together. But the third person you don't really know well and you do not like her very much. She is often absent from college and always hands assignments in late, or at the last minute, and does not work as hard as you or the other two group members.

- You are working in a group with three other people who usually get pass grades and are happy with their grades. You need a distinction on this assignment if you are to keep your grades up to the mark you want.

- You are part of a group doing a task that requires some research. One of the group members agreed to go to the library on his way home from college and bring the information into college today. But he is absent from college today.

How could some of these problems be avoided?

How to organise for group work

- Agree a timetable for action, including meetings to review progress.
- Agree on who does what in the group and make sure each person takes an equal share of the workload.
- Work to the best of your ability.
- Listen to other people when they are explaining their ideas.
- Sometimes when you are with friends, it is tempting to talk too much about other things, rather than the task in hand. If this happens, bring the group back to the task.
- Always try to plan for people being off ill/unable to get work done on time/ lazy/unable to find a book. It isn't always possible but you should try to get information early and have a few things on the go at once, so that you have something to do if things go wrong.

Learning on your induction

At the start of your BTEC programme, some time will be set aside to induct, or introduce, you to the main aspects of the course. Your induction will probably include most of the following for a Diploma student and some of them for a Certificate student:

- An ice-breaker exercise, like the one in Activity 2, to put you at your ease and to enable you to get to know the other students and staff.
- A visit to the main college facilities, such as the library, computer rooms, student union and student services rooms.
- Some group activities to give you experience of learning through doing.
- A thorough explanation of your course, probably leading to your class producing a briefing sheet of basic information as a learning through doing exercise.
- Some initial assessment, perhaps including a self-assessment, of your existing business knowledge and skills.
- A visit to a local business to see what it's like in a real business.

The purpose of the induction is to give you the basic 'survival kit' to start on the course. You find out what you are doing, why and how. You will also find out what to do if you have a problem. Remember, it is a learning programme, for as well as giving you the basic information you need (much of which could be sent to you by post before you start), you are also experiencing the BTEC style of learning through doing.

Activity 20 Your timetable for the year

Find out your timetable for the year. This should include:

- Your weekly timetable.

- When your work experience takes place.

- Term dates.

- Dates for your final assignments.

- Dates of any special events, such as residentials, visits to companies and so on.

Learning blocks

All the ways of learning have been listed separately. But a common way to organise a BTEC National programme, especially the core, is in blocks. This means that, for several weeks (the 'block'), you work on a theme or topic, usually with others. Within the block, the activities will draw from many, possibly even all, your core modules and some options too. During this time, you will carry out a number of activities, but these may not be assessed. When you get to the end of the block, you will finish the major activity, which will be the assessed assignment.

The end of a block is a good time to review your programme and update your action plan. In fact, an essential feature of the block is that at the end a tutor gives you feedback on your performance and looks forward with you to what you will do in the next block.

Activity 21 BTEC quiz

This section has introduced you to some of the fundamental ideas that you need to understand to be a successful student on the National. Working in small teams, draw up a list of the key terms to describe how you learn on the National, together with their meanings. Each team will take part in a quiz chaired by your tutor, who will draw up a list of all of the key terms identified by all of the groups. The quizmaster will ask each group in turn to define a key term. One point will be awarded for a correct answer. Another group can get two points if it can give a better answer. Some prize for the winning team will add spice to the quiz – if everyone puts some money, say 20p, into a 'pot' for the winning team, this will produce a bit of an incentive. At the end of the quiz, the class will produce a dictionary of all of the terms and their correct meanings.

Who can help you to learn?

If you are going to be successful on the National, it will be mainly because of your own efforts. You teach yourself as you work through assignments. You won't be spoon-fed or have someone standing over you telling you what to do all of the time. You will be given assignments or work to do and will be expected to get on with it. Nevertheless, the teaching staff are there to give you advice, to tell you how you are progressing and to help you plan to improve your performance.

The programme manager

The first point of contact for you is the programme manager – sometimes called the programme co-ordinator – who is the person in charge of your course. He/she will be aware of all of the facilities available in college and will be able to help you to get the best out of them. You will approach the programme manager first, in many cases, if you need help with any aspect of the course. If you are going to have to miss an assignment deadline for some good reason, the programme manager is the person to whom you should explain this. If you are going to be late into college for a few weeks because of an illness in the family; if you want some extra help with maths; if you want to know the dates of the terms and haven't kept a note of them; if you want a reference for a job; or if you want to change from full-time to part-time study after getting a job, then the programme manager is the first person to see. If he/she can't help, you will be referred to the person who can.

The class tutor

You will also probably have a class tutor, who is responsible for your class or group. You can approach your class tutor with any of the problems just mentioned. In a large college where there are several groups on the National, you may find that you get to know the class tutor far better than the programme manager, who may not even teach your group.

Module and common skills tutors

As the course consists of core, options and common skills, you are likely to have core tutors, option tutors and perhaps a common skills tutor.

Specialist tutors

In addition to staff who teach individual modules, you may find some members of staff with a specialist responsibility across the course. A lecturer may, for example, be responsible for Information Technology or Communication across the course, and so may come in to teach these particular topics on several modules. Get to know who these specialists are so you can approach them if you have problems in their areas.

Library staff

A modern college library is not just a place full of books which you can consult in silence. Increasingly, college libraries have a whole range of materials to help you learn. As well as books and magazines, a typical college library will have Prestel/Teletext, audio and video recorders, a library of audio and video tapes, and computers for you to use. There will often be special seminar rooms that can be used by groups of students who need to do research or project work in the library. Unit 8 discusses library resources in more detail.

The library staff are the people who know most about what is in the library and where to find anything you need. They can help you to find information, if it's in the library; call books back from other borrowers if you need them; borrow books or journals from other libraries for you; keep clippings files on major topics, which are useful for project work; and show you how to use the equipment. In return, they ask that you follow the rules of the library – not a lot really for what they provide.

Careers tutors

The lecturers are not just there to help you with your course but also to help you to move smoothly on to a job or further qualifications after the National. Your class tutor will give you advice on what to do after the National, but your college probably also has specialist careers tutors. The careers tutors in the college can do a number of things for you. They will have information on the different colleges and universities that offer degrees or HNDs that you might be interested in taking when you have finished the National. They'll be able to suggest courses to you, and tell you about them and what grades you'll need to get on to them. When you've decided what you want to do, they'll be able to help you with the paperwork, such as filling in grant applications and university, polytechnic and college application forms (often shortened to UCCA and PCAS), and tell you about grants, accommodation and the colleges that you're interested in.

For Certificate students, the careers tutors and your class tutor should be able to answer all of your questions if you want to go on to a part-time HNC or degree at your local university or polytechnic.

If you're a full-time student looking for a job straight after the National, they'll be able to help you to look for one and to suggest ways of improving your chances of getting the jobs that you apply for, perhaps giving you a mock interview. They may also be able to suggest careers that you hadn't thought of or give you information about the ones that you have thought of, to help you to decide what is right for you. They'll also be able to tell you about the qualifications that you'll need to get in order to succeed in different jobs and how to go about getting them. The careers tutors work closely with the government careers service and can arrange interviews for you with a careers officer to give you more advice on choosing a career.

Every college provides a careers and higher education service, although it may be called 'student services' or something similar. The service tends to be used mainly by full-time students, but it's available to part-time students needing confidential and impartial advice, as well as information on further qualifications and job opportunities.

Counsellors

Most colleges have trained counsellors to help with all sorts of personal problems that you may come across during your time at college. They provide advice, which is strictly confidential and will not be passed on to parents or employers, on how to claim benefits, health and personal problems, accommodation and any other problem that you might have. So, if you have a problem that you need to discuss with someone in confidence, you will find that the college counselling service will listen and, if they can, help.

They don't just deal with personal problems. They can also help with things like a fear of exams or arrange help with poor spelling or maths. If the college counsellors can't help you themselves, they will be able to put you in touch with someone who can, or contact them for you if you prefer.

Activity 22 'Who's who' of staff

To make the best use of staff, you need to know 'who's who' – who is responsible for what and what each person can do for you. You also need to know where you can find each person.

Draw up an organisation chart or family tree of all of the staff who are responsible for the different aspects of your course, as well as those in the college who can be useful to you, such as library or counselling staff. Indicate their location and any other relevant information – if, for example, a member of staff works on another college site on certain days and can't be contacted by you.

Your chart will vary according to your college but will probably look something like that shown in Figure 3, although this example is only in outline and your own chart will have more people on it.

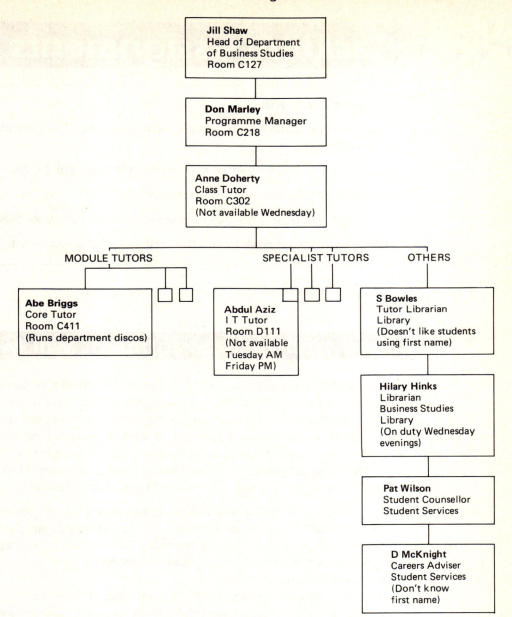

Figure 3   Organisation chart of 'who's who'

UNIT 4 Tackling assignments

In this unit, you will:

- learn what assignments are and what part they will play in your programme;

- identify the main types of assignment that you will be given;

- find out how assignments are presented;

- identify the main activities in assignments and on work placement;

- discover tips on how to get the best out of the work that you do.

And your next assignment will be '...'

This sounds like something that you'd hear in a James Bond film doesn't it? You'll also hear it a lot on your course. The assignments that you'll get will be a bit different of course, but the principle is the same – an assignment is a job that you've got to get done. To do it well, you need to use all of the knowledge and skills that you've gained, and make full use of all of the resources available to you.

So what is an assignment? You've already been given some information about assignments in Units 2 and 3. You already know that assignments:

- Are business problems which you tackle and which you are assessed on.
- Are not simply 'tests' to see how well you have learned something, but ways for you to learn. You'll learn by preparing for the assignment and also from the feedback that you receive on your performance.
- Are practical business tasks.
- Determine the grades that you get for each module at the end.
- Are opportunities to prove you have gained particular skill or module elements.

This means that assignments are a very important part of the course and a major means by which you'll learn. They are at the very heart of the BTEC style of learning through doing. You'll learn when you are completing the assignment and also when your assignment has been marked by your tutor, who will spend time with you discussing how well you did and how you can improve.

What types of assignment are there?

Throughout your course, you will have different types of assignment to do.

Module assignments

These assignments are aimed at helping you with a particular core or option module. They don't involve much knowledge from other modules.

Integrative assignments

These assignments involve two or more modules and try to draw together knowledge and skills from a few different areas, to help you to see the way that the real business world operates.

Work-based assignments

You do these assignments when at work or on your work experience placement. You'll find a special section on these assignments in this unit, because they are so important to the course.

Community-based assignments

These assignments involve you going out of college into the local community to investigate and carry out tasks. You may go into a shopping centre to carry out a survey or into a particular neighbourhood to find out about social conditions. Wherever your assignment takes you, the skills and knowledge that you have acquired will be applied to a real local problem or issue.

Final assignments

These assignments are set at the end of the module and may count more in your final grade than earlier assignments. They may also cover a wider range of topics than normal assignments. Don't confuse them with traditional exams – they will only be part of your assessment and will be similar in style to the assignments that you have done throughout the year.

Assignments can also be further split into:

• **Assessed assignments** – assignments that are graded and contribute to your overall results.

• **Non-assessed assignments** – assignments that are not graded and do not contribute to your overall results, but which give you useful experience of tackling business problems.

Activity 23 Your assignment programme

Find out how many of each type of assignment you will have throughout your first year in term 1, 2 and 3. Write these out on a calendar to help you to plan your year and to know what is coming up in the way of work at any time.

Standard layout of assignments

Each assignment that you get will be different, although they all have some things in common. You will find lots of detailed assignments in Units 6 and 7 of the book and, if you have a look at those assignments, especially the ones in Unit 6, you will see that there is a standard layout for the assignments. Your college may use a different layout to the one used here for assignments, but you should find that they give you the information you will need. Common sections in an assignment include the following.

Introduction

This gives you a bit of background information, gives you some idea of the information that you will need to find and where you might make a start in looking for it. You are told which of the skills you will be required to use in the assignment, together with some idea of how important each of them is.

Elements covered

A brief list of the modules and outcomes the assignment covers.

Assessment criteria

These tell you what is expected for a pass grade and for a distinction grade. A merit grade would fall between the two. The assessment criteria don't tell you the answer to the assignment, but they do make it very plain what you have to do to get the different grades.

Situation

This explains the background to the assignment and the situation that you're in, as illustrated in the following example.

You are the recently promoted manager of a small branch of a national company. You are younger than many of the people who work for you and feel that they do not respect you. This morning, one of the supervisors arrived late for the sixth time in two months. You mentioned the matter to him and he became abusive, telling you that he wasn't '... taking orders or correction from someone barely out of nappies ...'. He then said that if you tried to discipline him '... you'll have a riot on your hands; I've been here for years and they'll all back me up'.

Some situations explain what you have to do, while others could 'set the scene' for a role play, request a report for the managing director of a business or request that you carry out a survey.

Tasks

This is the actual work that you will have to do and is broken up into sections. Some of these sections will require individual work, while others may necessitate you working with other people. A single assignment can be made up of quite a lot of different tasks. These tasks could involve you in doing one or more of the activities described in the next section.

Grades and comments

This provides a record of the grades that you are given for the assignment, together with any comments from your tutor. If you want any more feedback, then go and see whoever marked the work and have a chat about what you did and how you could have done better.

Tackling different types of assignment activities

For most assignments, you will have to carry out a number of activities; for example, research information on a particular firm, or carry out a survey or interview people and make a written or verbal report on your findings. To be successful in your assignment, you need to be able to perform well in all of these activities. In an assignment, as in a job, it's no use being good at only part of your work. You need to be on top of all aspects of your work.

This section lists the most common assignment activities you will come across and gives you tips on how to tackle them. You will find it useful to keep referring to these tips as you progress through your course.

Case studies

What

A case study gives you detailed information on a real or hypothetical business situation or problem (the 'case'), and you have to work out what the answer is or what action should be taken. You are dealing with realistic business problems that have been simplified so that you can learn from them and tackle them successfully.

Examples

- A company wants to build a new factory. You are asked to recommend a new site from the information provided on three alternative sites.
- A company is launching a new product. You are asked to design a promotional campaign targeted at a specific category of customer.
- A new office supervisor has antagonised long-serving office staff who are unhappy and demotivated. From the information provided, you are asked to explain the causes of the problem and what could be done to overcome them.

Tips

- Read all of the information in the case study and ask your tutor for clarification on anything you are not sure about.
- Make sure that you know exactly what problem you are being asked to comment on in the case study.
- Plan your time to gather any information you need so that you can meet the assignment deadline.
- Make sure that your answer is in the required format (letter, memo, oral presentation) and is addressed to the right people, using the right language.
- Make sure that your answers relate to the problem in the case study. It's a common mistake to give a generalised answer that doesn't deal with the specifics of the case study.

Role plays

What

A role play involves you in acting or playing the part (the 'role') of someone else in a business situation. Often, fellow students, tutors or business people play other roles. As with a written case study, you are dealing with realistic business problems, but you are now part of the action. Often, there is no correct answer to the business problem – an answer evolves during the role play depending on the interplay of the various players.

Taken seriously, a role play gives you the chance to experience all sorts of situations that you may meet in business, such as chairing a meeting, interviewing for staff, presenting a case to the Board or selling to a client. This practice is valuable as it allows you to make mistakes at college rather than at work – losing control of a meeting at college and understanding why is a very valuable lesson if it means that you won't do the same at work. Role-play exercises are often video recorded so you can look back afterwards and see how you and the other players performed.

Role play is one of the most powerful ways for learning about business, but to be successful all of the players have to enter into the spirit of the exercise. Its special value to you is that it puts you in an unfamiliar role and, in playing the role, you not only learn how to handle business problems but also how to handle people. In a really useful role play, you can also learn a lot about yourself and how other people see you. Role plays of job interviews are very important in this respect.

Examples

- In a negotiation over a wage claim, you have to play the role of a trade union representative.
- In the role of a receptionist, you have to deal with a phone call from an awkward customer.
- In an appointment interview, you have to play the role of one of the appointment panel, interviewing students who are playing the roles of the interviewees.

Tips

- Read your own role very carefully and ask your tutor for clarification of any points you don't understand.
- Try to work out how the person you are playing would feel and act in the situation.
- Be prepared for action – work out your tactics for the role play, especially if your role play involves potential for conflict with other role players. If so, try to guess their likely tactics.
- Use the kind of language and gestures suitable for your role.
- Be flexible during the role play. Be prepared for the other role players to surprise you.

- Keep calm and logical whatever your role and however heated the role play might become. Remember, it's only a game!
- If the role play is recorded on video, watch it carefully afterwards and be honest with yourself on how you performed and how you could improve.

Researching information

What

You will often have to gather essential information that is needed to tackle a business problem. Most assignments will involve an element of researching information, such as consulting a textbook or newspaper. In some cases, however, where information is not readily available, then uncovering and presenting it becomes a major element in the assignment.

Examples

- A company wishes to install a Fax machine. You have to discover the most suitable from those on the market.
- A company sells financial services to people aged over 50. You are asked to research population trends for this age group for the rest of the century.
- A company is considering entering a new market. You are required to report on the main trends in ownership, and the control and profits of companies already established in the market.

Tips

- Make sure that you know the main sources of information in your college and locality (Unit 8 deals with these in detail).
- Give yourself plenty of time to collect your information – it's usually very time consuming.
- Set yourself a timetable for collecting and analysing the information. Be prepared for frustrations, as information is often not available in the way that you need it to be.
- Remember that your research is only useful in business if it is clearly presented in a businesslike way.

Surveys and questionnaires

What

Where information is not available in print, firms often set about collecting their own information by asking questions directly of the people who matter – their customers, staff or the general public. During your course, you will be told what information is required and you will have to collect that information. The questionnaire is the most common and cost-effective way of surveying a large number of people, and is used widely in business.

Examples

- A company wishes to restrict smoking on its premises. You are asked to design a questionnaire to discover the views of employees.
- Your mini-business is establishing a college shop. To ensure it has the right stock, you have to design and administer a questionnaire to find out what kinds of good the students want.
- Your local council wants to survey bus passengers to find out what they think of the services. You have to design a questionnaire and administer it to passengers during their journey.

Tips

- Try out your questionnaire on a friend before using it in the assignment.
- Make sure that the questions are clear and unambiguous.
- Make sure that you can explain the meaning of each question in your own words before using it.
- When interviewing people for the survey, always introduce yourself as a student and explain the reason for the survey.
- Always be polite and do not pressure people who do not want to take part in the survey. Remember, they may be in a hurry.
- Be prepared for some people ignoring you or being rude to you. Most people are co-operative and friendly towards students conducting surveys. But you can't win them all – don't be discouraged by the rejections. There is always someone else to ask.

Visiting local organisations

What

You will regularly make visits to local organisations, sometimes on your own, but more often with a group of students and a tutor. You may be required to make some of the arrangements for the visit.

Examples

- A visit to a local firm to see a new computer system.
- A tour of a chemical plant to see production methods.
- A visit to the local Town Hall for a tour and a talk on services provided to businesses and the community.

Tips

When organising the visit:

- Try to give yourself at least three weeks to plan.
- Write down the reasons for your visit – what exactly do you want to find out about?
- Agree among the group who will do what to organise the visit, and put it in writing with deadline dates for jobs to be done.
- Agree with the firm on visiting times, purpose, what areas of the firm will be seen, and any safety or security requirements. Agree a meeting place at the organisation.
- Plan questions before the visit and send a list to the organisation with confirmation of arrangements for the visit.

During the visit:

- Give yourself plenty of time to get there, if you haven't been to the organisation before.
- Dress appropriately – you may apply for a job there some time.
- Turn up on time, at the agreed meeting place.
- Introduce yourself and the group.
- Take notes during the tour (a clipboard is useful).
- Ask questions, but be polite, and remember that anything heard and seen should be treated in confidence.
- Thank the people who helped you on the visit at the end.

Afterwards:

- Write up your conclusions and findings straight away, while they are fresh in your mind.
- Send a letter of thanks.

Organising events

What

As part of your course, you and other students may be asked to organise an event under staff supervision. The event may not seem very relevant to business, but in taking responsibility for organising something, you gain valuable experience for organising yourself at work.

Examples

- A day trip to the House of Commons for students.
- A charity fund-raising event.
- An evening reception and exhibition of student work at college for employers and managing agents.
- A seminar for local employers on a topic of current interest, such as developments in the European Market.

Tips

- Give yourself plenty of time to plan the event.
- Draw up a list of jobs and deadlines for their completion.
- Make sure that everyone in the group has a job to do.
- Write down the jobs and their deadlines as a 'contract' for each person to fulfil.
- Make sure that the materials needed for your event are available, such as cash, rooms, outside speakers, transport and so on.
- Draw up a promotional campaign to publicise the event and, if necessary, to attract people to join in.
- Hold regular meetings to check on the progress and that everyone is fulfilling their contract.
- Have a trial run of the event shortly before it is due to take place, to iron out any problems.
- Make sure that everyone has a job to do on the day of the event.
- Make sure that everyone turns up with plenty of time to spare before the event starts.

Other practical activities

What

The aim of your course is to make you competent to work in a company, whether as a receptionist, secretary, clerk, trainee accountant, marketing assistant and many more jobs. Each job has its own specialist equipment, documents or procedures, which you need to be on top of. Consequently, you may be required to display highly specialist, practical skills during your course.

Examples

The list is endless, since each job has its own practical skills, which are not used elsewhere. Typical examples are:

- A secretary will have to develop shorthand or audio typing and advanced keyboarding.
- Someone working in a finance office will have to deal with order forms, invoices and a host of specialist financial documents.
- Someone in a salaries section will need to be able to handle timesheets, and calculate wages and bonuses.
- Someone in a marketing office may be involved in organising promotional events, including booking hotels and accommodation.
- Someone in a purchasing office will need to be able to handle a lot of specialist order forms and perhaps receive and record goods when they arrive from a supplier.
- Someone working in a supply office may have to complete crucial documents to ensure that far-flung parts of the company have the basic supplies and equipment needed to continue production.

Every job has its specific requirements for practical skills. Knowing what to do is not enough. You have to be able to do it, and your assignments should give you a good preparation for performing well in any job.

Work-based assignments

In Unit 3, you will have noted the importance of work experience for full-time students and of the workplace for Certificate students. Since there are advantages in both work experience in an organisation that you aren't familiar with, and in using your daily work to gain information and insights for the course, you'll see that there are sections here on both.

Work experience for full-time students

Work experience may not be new to you – you may have a part-time job, have worked over holidays or been on work experience from school. The way that it will be used on this course may, however, be a bit different, because you will have assignments to do during the time that you spend on work experience, as well as before you go and after you come back. All of these may be assessed, and all of them aim to help you to get the most out of the work that you are going to do.

Examples

- Find your own placement.
- Visit the company for an interview before being offered a place.
- Prepare a curriculum vitae (CV) for the firm before placement.
- Keep a log or diary of what you do and learn while on placement.
- Draw up a short report on the company: its legal status, structure, main products/services and clients.
- Carry out a mini-project for the firm on a topic that it suggests.
- Prepare a report on the work experience and give a presentation on it when you return to college.
- Compile evidence of work you have done.

Tips

You will find a lot of tips on how to tackle different types of assignment on work experience in other sections of this unit. The section on organising a visit, for example, is especially useful for work experience.

A number of points to remember on how to get the best out of the work placement are:

- Remember that you are a guest of the company. You are going to get more out of the time that you spend there than it will.
- Find out about the firm before you go, so that you can be seen to have taken an interest in it.
- Remember that you will be treated like an employee for the time that you are with the firm. Dress appropriately, keep good time and follow the rules of the company, even if you don't see the point of them.
- At the end of each day, make a note of what you have seen and done and how it ties in with what you have done at college.
- Be inquisitive and ask questions about the firm and the work that you and others are doing. You are there to get a picture of the business and the way that the business world works, so make the most of the time.
- Remember that the people that you are working with have their own jobs to do and, while they will help you as much as they can, there will be times when you need to let them get on with their work, without interrupting them.
- Anything that you hear, see or find out is confidential. Firms don't want their secrets or scandals spread around the town or the college. Always get their approval before using any documents or information in your assignments.

- There will also be a lot of non-confidential information around. Make the most of this and collect things that may be useful to you, such as rule books, publicity materials and forms. But remember to ask permission first.

Work-based assignments for part-time students

You will be required to relate many of your assignments to your workplace. Your assignments may require information on your job, on the industry or about the way that the firm works. This will often require the co-operation of a supervisor or training officer, who may also be involved in the assessment of your assignments.

Examples

- Draw up an induction programme for new recruits joining the organisation.
- Design publicity material for one of the firm's products or services.
- Report on the extent that the organisation uses new technology and identify areas where it could be applied with advantage.
- Discuss with others in the organisation what their jobs include and what they see as the easiest/most difficult jobs that they have to do.

Tips

The tips given in other sections of this unit, especially the sections on researching information and designing questionnaires, also apply to many work-based assignments. There are some specific tips however:

- Enlist the support of your colleagues and supervisors whenever possible.
- Make sure that the organisation is aware of, and doesn't object to, the assignments that you have to do.
- Look for allies – make contacts with people in the firm who might be able to help you, such as training staff, senior managers in the area that you work in, and people who have done the same course in the past or are doing a similar one at the moment.
- Collect any non-confidential information and documents that you feel could be of use to you on the course. Don't wait until you need something and then try to remember who has a copy. When you see something useful, ask if you can have a copy there and then, and store it until it's needed.
- If you're going to use any information outside the firm, make sure that you have permission to do so.

- If an assignment takes you into an area that the company is sensitive about, make sure that your tutor is aware of this, and make sure that the firm is happy with what you have done, before handing it in.
- Be honest with yourself and your tutor when discussing your firm. It's easy to be over-critical. It's also easy to end up with an assignment that reads like a press release and ignores any difficulties.
- Try to make sure that the work you submit is your own, rather than a 'cut-and-paste' job that is based on the company documents and what colleagues have said.

Presenting your answers to assignments

When you do assignments, you will be asked to present the answers in certain ways. The following sections look at some of the most common ways.

Presenting information in writing

Writing is one of the main ways that people communicate within organisations and how organisations communicate with each other. Communication is, therefore, one of the seven main skill areas on your course, and is assessed through all of your work. It makes sense, therefore, to ask you to present your answers to assignment tasks as you would in a firm or business. This gives you practice in writing business documents, and you will be assessed on the style and layout of the document as well as the content. In other words, you are assessed on how good you are at business communication. Depending on how well you do, you may want to put this into your action plan as something that you need to work on further to improve. The main types of written communication that you will use in business are reports, letters and memoranda.

Writing reports

A written report is an account of work that has been done on a particular task. Reports are used when a full and extensive account is needed – a simple letter or memo won't suffice. They are also very formal, in that there are some general rules to be followed. (Some reports are informal, but these still have rules that you have to follow in presenting them.)

Examples

- A customer survey may have been carried out and you are asked to write a report explaining the findings.
- A director may want a report on the work done by a department during the year.
- Your manager could ask you to produce a report on the possible consequences of starting a new product line.
- You could be asked to outline the law relating to data protection as it affects your section in a report for your supervisor.

Tips

- Although there are no hard and fast rules for reports, they are usually divided up as follows:
  - A **title** page, which gives the title of the report, the name(s) of the author(s) and the terms of reference, as illustrated by the following:

    A report on the laser printers suitable for the legal section compiled by Ms P. Nathubai, Legal Assistant, Legal Services Section

    Terms of reference: On 13 January, Ms P. Nathubai, Legal Assistant, was asked by Mr Tamburello, Head of Legal Services, to investigate the costs and facilities of laser printers and to identify those that would suit the needs of the legal section.

  - A **contents** page, which identifies the sections of the report and gives page numbers.
  - An **introduction**, which explains what the report is about and the reasons for doing it.
  - A **procedure** section, which explains what the author(s) did in order to get the information to compile the report.
  - The **findings** of the report. This will form the biggest part of the report, explaining what the authors have found out. This part of the report should be divided into sections, with each section referring to a separate point. The sections should be numbered so that people can refer to them easily.
  - The **conclusions** of the report, which state what the authors found out in brief, as illustrated in the following:

    It is the opinion of the writer that the legal section will be unable, for the reasons stated in the findings of the report, to purchase a suitable printer under £3000 and that there will also have to be an investment of £1200 in software. The following models on the market meet the requirements of the company ... .

- The **recommendations**, which outline the action to be taken.
- The **appendices**, which set out information that has been referred to in the report and which could be useful to someone who wants to see the detailed evidence used in the report, but is not necessary in the main body of the report.
- The **references/sources**, which list the books or references referred to in compiling the report. Each reference should be referred to by title, author, publisher and year, as in *Social Trends*, HMSO, HMSO Publications, 1991.
- The report should be divided into sections, with each section having a heading and a number.
- A report has to be written in the third person. This means that you shouldn't use 'I have looked at the possibilities ...'. Instead, you would say 'The various possibilities having been investigated ...' or 'The writer investigated the various possibilities and ...'.
- An informal report lets you use 'I'.
- You may leave a wide margin after the title in each section. Wide margins are used in reports so that there is space for readers to make notes as they go through them.

Writing letters

Letters are frequently used in business. They can be used within a firm but are more often used to communicate with people outside the firm – for example, other firms. Letters can deal with quite a lot of information and can be used in all sorts of different situations. An important reason for using a letter is that it gives you a permanent record of what is said, so that there cannot be any disagreement later. So, even when a communication is carried out face to face or by telephone, it is often followed up by a letter of confirmation, and a copy is placed on file.

Examples

- Letters are sent to suppliers to ask for information.
- Firms write letters of complaint to suppliers.
- Firms send letters requesting payment to customers.
- References for employees are often in the form of letters.
- Disciplinary warnings are sent by letter.

Tips

- Letters start with a **salutation** and end with a **close**. There are two possibilities:
  - Dear Sir/Madam ... Yours faithfully
  - Dear Ms/Mr ... Yours sincerely.
- Most firms now tend to use a 'fully blocked' format for typed or word-processed letters. This means that all of the text of the letter starts at the left-hand margin and there is no punctuation, except in the main body of the letter itself.
- Provide both your address and the address of the person you are writing to.
- Put the date on the letter, together with any references you may have for it.
- Print your name under the space left for your signature – signatures are often illegible.
- Where there are documents to go with the letter, you should write 'enc.' on the bottom of the letter and list the documents enclosed, so that the person that you are writing to knows that documents should have been enclosed with the letter.
- Try to keep sentences short.
- Don't try to make more than one point at a time.
- Avoid using complicated language, where possible.
- If the person writing to you has used a reference number for his/her letter, quote it back on yours. This will help him/her to locate the file quickly and easily.

The sample letter in Figure 4 incorporates these tips, but in Activity 24 you have to show that letter writing isn't just about following rules of style and layout. You also have to get the language and tone right.

Activity 24 'I regret to inform you that ...'

As part of an assignment on staff selection, students were asked to write a letter to candidates who didn't get the job of a clerk in the Accounts Department of Chapnor Services Ltd. Remember that any letter sent out creates an image of the whole business, as well as of the person who sends it.

1 If you received the letter outlined in Figure 5, what image would you form of the company and the person who sent it?

2 Read the letter carefully and identify any faults – for example, tone, content and layout.

3 Rewrite the letter so that it reads better and could be used as a business document.

DIRECT CAR SALES LIMITED
Registered in the UK No:145983

Registered Office:
Coverdale Cottage
The Green
Sinnington
York
YT6 8UN

Mr Jones
45 Sunningdale Drive
Thornton
Lancashire

21–4–92

Our Ref: JN/34.01

Dear Mr Jones

Re: Supply of Granada Scorpio

I have pleasure in informing you that the car that you ordered is now ready for collection from the supplier and will be delivered to you on Friday of this week. As requested, we have arranged for the installation of a sunroof and full towbar assembly on the car.

There is no balance outstanding, your cheque of last week having covered the cost in full. Please ensure that you have all of the keys to your part-exchange vehicle with you on Friday so that the delivery driver, who will take the vehicle away, can pass them on to us.

Thank you for your custom. I look forward to dealing with you again in the future and wish you trouble-free motoring. Assuring you of our best attention at all times.

Yours sincerely

J. Tudor

J Tudor (Director)

enc: Copy invoice
    Receipt
    Brochure

Figure 4 Sample letter

CHAPNOR SERVICES LTD.
85 SHAWLEY GATE,
BRADFORD

Mr. Jason Wallburt,
18 Birch Lane,
Bradford.

Dear Mr. Walburt,

I am sorry to tell you that you have not been appointed as Accounts Clerk. I hope you are more successful in the future.

Yours sincerely,

N. Levett (Mrs)
Personnel Department

Figure 5   Letter for Activity 24

You may have noticed that the letter is missing the following details: the date, a reference number and the telephone number of the organisation. In addition, the candidate's name has been written incorrectly and the tone is very wrong. And Mrs Levett has not troubled to sign it.

Writing memoranda

A memorandum or 'memo' is used to communicate with other people in the organisation, which is in contrast to a letter, which is more often sent to people outside the organisation. Memos are cheaper and quicker to write than letters, and thus save money for the firm. A memo provides a quick way of giving information or asking a question.

Examples

- Meetings are often notified in a memo.
- Memos are used as reminders of matters that people may have forgotten.
- A memo could be used to ask a short question or request some information.

Tips

- Remember the KISS rule: **K**eep **I**t **S**hort and **S**imple.
- Make one or two points only in a memo. Try not to make it too long.
- You don't have to sign a memo.
- You can often find pre-printed memo forms, which only need the fine details added, to save you time.

Activity 25 *'I'm sorry but ...'*

You have received the memo shown in Figure 6. Unfortunately, you have another meeting from 9.00 to 10.30 AM, which you definitely cannot cancel. Write a memo to Alan Brown telling him that you cannot attend his meeting until 10.30 AM.

M E M O R A N D U M

From: Alan Brown   Date: 23-4-92

To: All business managers

Re: Sales meeting for April

Please could you attend a meeting on the 29th of this month at 10 AM in the main boardroom to discuss this month's sales figures. Please bring your department cost figures for the month to the meeting. I would hope that two hours will be sufficient.

Agenda to follow.

Figure 6   Memo for Activity 25

Giving presentations

A presentation is the term that is used in business for an event that is used to present information to an audience. However, presentations are not just ways of giving information, they are ways of persuading people to accept the point of view of the presenter. So, presentations need to look and sound professional and give all of the necessary information. On your course, you will be asked to present some information on your own or as part of a group. In either case, you will have to consider how to get what you want to say across to your audience in a simple, interesting and persuasive way. Audio-visual aids, such as flip charts, projectors, overhead projectors and video recorders, will be a great help to you in making presentations, and it is likely that you will need to use some of these during the course.

Examples

- An advertising agency would give a presentation to show how it would handle a product to try to get a big account.
- An engineering firm would give a presentation to clients to show how its product would help the clients' business.
- A department head in a firm might use a presentation to explain to staff in the department about a new product line to be introduced.

Tips

- Decide the aim of your presentation.
- Plan your presentation in advance.
- The presentation should have a clear structure – an introduction, a main body and a conclusion, and your audience should be aware of what you intend to do.
- Make notes to speak from but only use them as a prompt – do not simply read the notes, as this is the surest way to lose the audience's interest.
- Use language that is suitable for your audience.
- Use written summaries, video, overhead projectors or other aids to illustrate your talk and keep the interest of your audience.
- Introduce yourself to your audience, if they do not know you.
- Tell your audience when questions can be asked – you may want to leave time for questions at the end or you may be happy to answer questions at any time throughout the talk, but interruptions can sometimes lose the audience's interest and disrupt the flow of your presentation.
- Summarise the main points at the end and ask your audience if there are any further questions.

Other ways of presenting your assignment

The methods already outlined for presenting your assignments are the ones that you will use the most. But there are others that can be used. In fact, an assignment can be presented in any way that the tutors feel is useful and could be used in business. It isn't possible to cover all of these other ways of presenting your assignments, so the following sections just look at some of the more common ones, with some advice on how to use them.

Attending an interview

There are quite a lot of situations in business where you could have to attend an interview as part of your assignment. Interviews can also be used as a means of assessing your work on an assignment. When you have to attend an interview, there will often be some written work to do as well.

Examples

- You have completed an assignment on interviewing and employment law and have to conduct interviews for jobs in accordance with the law.
- You have to be interviewed by a panel of tutors and people from business on the findings of a report that you have produced.
- You have to carry out disciplinary or appraisal interviews for staff.

Tips

- Make sure that you dress appropriately, if you have the chance.
- Remember that you have only one chance to get things right in an interview. So, think before you answer questions or ask them.
- Do your preparation before the interview and try to think of the questions that you could be asked. But avoid seeming over-prepared, by answering questions with set answers that you have memorised (and sound like it).

Producing a video

Videos are often used in business to get information across in a vivid way. However, they take a great deal of skill to prepare really well. Videos have the advantage that you can take your time to set up what you want to say and use the video to get it across quickly. Also, you can always 'redo' a section of a video.

Examples

- Prepare a video to explain the annual report and accounts to the staff of a business.
- Make a short film to explain to new recruits in the company how the firm is organised and what they do.
- Prepare a training video for the personnel department to show the right and the wrong way to deal with lateness.
- Make a short video to promote a product or service.

Tips

- Don't try to imitate fancy special effects if the system isn't designed for it.
- Find out what sort of facilities the system that you are using has – 'fades', slow motion, windows and colour changes can be very effective if not over-used.
- Make sure of what you want to do by planning the video carefully.
- Avoid 'talking heads' type of films, which are often boring and could be better done live.
- Allow time to review what you have done and to change it if it looks wrong.

Handing the assignment in

When you present your final answer, it is important for it to look businesslike. In addition, it should be accurate and well researched. Good presentation isn't a substitute for doing the work well, but it can help to make the material easier to read, and, therefore, whoever is reading it will get more from it.

Tips

- It is a good idea to use a new page for each task and to label them clearly, so that they can be found quickly.
- Check your work for errors in spelling and layout. Make sure that there are no crossing-out marks and so on.
- You will find that your work will look better if it is word processed or typed. That said, neat writing is always acceptable and may be better than poor typing.
- Put the completed assignment in a folder. (You don't need a new folder for each assignment – you can use the same one many times.)
- Work looks better if it has a neat and attractive front cover with the assignment title, your name and the name of your tutor on it.

- Provide a contents page so that people can see what you have done, and label any tables, charts or diagrams clearly so that they can be identified from the contents page.
- Include a bibliography if you have referred to any books.
- If you have made visits, written to organisations, made telephone calls and so on, it helps, when your tutor is assessing your work, if you have details of these in your assignment, even if they were not asked for.
- Put any supplementary material in appendices at the back of the assignment. Don't try to use it to 'pad out' a thin piece of work.
- Don't confuse quantity with quality. Assignments can be too long as well as too short, and there are no points to be gained from using eight long words where one short one will do. The chairman of one of the biggest companies in the country insists that all of the information that he is sent should be on one side of A4. If he wants more information, he'll ask for it. That's probably too short for most of your assignments, but you get the idea.
- Always review your work before handing it in – it's easy to forget a section or not to have given enough detail somewhere.
- Finally, ask yourself 'Would I give this to my boss at work?' If the answer is 'No', then it isn't good enough for college either.

UNIT 5 | Assessment

In this unit, you will:

- learn what is assessed;
- find out how you will be assessed;
- look at the role of exams;
- identify who assesses you;
- learn how your grades are recorded;
- see how the final grades are decided;
- find out how to deal with problems that affect your grades.

In earlier units, you have come across quite a lot about assessment – how you'll be assessed and why it's important for you to know what's going on. This unit will go over some of what you know already but in more detail. At the end of this unit, you should have a clear idea of how you're assessed and how the grades that you get are used to determine your final grades.

What do you know already?

A lot has been said about assessment in earlier units – in fact, assessment is mentioned in almost all of the earlier units. This is because it's important that you understand how you're assessed. It's a bit like knowing the Highway Code when you're learning to drive – if you don't understand the signs, then you're not going to be able to follow them. What have you found out so far?

- You know that you are assessed through assignments. The assignments that you do are graded and these grades help to tell you how well you are doing, what standards you are achieving and where you need to do some work to improve (Unit 4 covers a lot of this).
- You know that you're assessed on all of the modules on the course, both core and option (see Unit 2).
- You know that you are assessed on your common skills as well as your knowledge (Unit 3 covers this).
- You know that you are assessed continuously and not just by an exam at the end, as happens with courses such as A-levels (see Unit 2).
- You know that it is important to review your work before you hand it in and after it has been graded, to make sure that you have done what was asked and to see if you can improve (see Unit 4).
- You know that at the end of each year you are given a grade for each module and one for common skills (see Unit 2).

What is assessed?

Your course consists of two main elements, both of which have to be assessed:

1 The modules, which provide the knowledge and skills that you need to deal with specific aspects of working in business, such as Business Law, or the core modules. All of the modules are divided into outcomes which set out what you actually have to be able to do. You have to show that you can meet all of the outcomes in a module in order to get a pass and it is up to you to keep track of what you have and have not done so you can claim the outcomes from the lecturers. The section on 'evidence' later in this unit will tell you more about how to do this.
2 The common skills, which are those seven vital abilities required for any job in business and which should appear right across the course.

This reflects the two aspects of any job:

1 The knowledge you need to do the job.
2 The skills you need to use the knowledge.

For example, if you were dealing with a customer's complaint, you would need to have the legal and technical knowledge relating to the complaint, to be able to do anything about it. You would also need the skill to apply the knowledge and the ability to handle people, to deal with the complaint effectively.

The knowledge is useless without the skills to make it work for you, and the skills aren't much use to you if you don't know what you're doing – so you need to be on top of both of them. The skills that you need to put the knowledge to work for you are the seven common skills. What you need to remember is that you will need to use these skills in almost any job – they move with you and they help you to move easily from one job to another. Being able to say that you work well with other people, that you can use information technology and that you can solve problems are assets to you wherever you work. They can be called **portable** skills, because you will carry them with you whatever work you do and use them in any business situation that you are in. For example, although you might not need a detailed knowledge of finance in every job, you will certainly need to be able to work with people and to communicate well.

The seven common skill areas each have a number of elements. In order to get a pass in a common skills area, you have to show that you can meet all of the elements in it. In most cases, this will be assessed by some sort of profiling system whereby you 'claim' skill elements when you feel that you have enough evidence to show that you can meet all of the requirements for it. This evidence may come from the work you do on your course or from experiences you had before, such as jobs or school. There is a section on putting this evidence together later in this unit.

Activity 26 Self-audit of your skills

At the start of your course, you ought to assess, with your tutor, your strengths and weaknesses in each of the common skills. You will probably find that you have a lot of previous experience that gives you a head start in many of them. Having a computer at home would give you a head start on the information processing skills. Some of the areas where your experience will help are less obvious: if you've helped to run a football team, worked in a shop part-time or done some voluntary work, then you will have developed some important skills, such as working with others and solving problems.

You could assess your skills on your own, but it's better to discuss your skills with three or four other people who know you well. For each skill area shown in Table 11, list all of your past experiences that you think are relevant. Don't just think of where you have studied the skills, such as in Maths or English at school, but also any work, leisure or domestic experience that has given you a chance to develop and use the skills. When you have worked out where you have experience, then assess how good you are in each area. Be honest with yourself – knowing your strengths and weaknesses is the starting point for improvement.

TABLE 11   *Assessing your skills*

Skill	Experience	Level		
		Need help	OK	Good
1 Managing yourself				
2 Working with others				
3 Communicating				
4 Solving problems				
5 Numeracy				
6 Using technology				
7 Design and creativity				

Activity 27 Skills for your job

For a part-time student, it is especially important to assess the skills required for your job. Do you have to work closely with others, use a computer for data processing, have a high level of numerical skills? Different jobs will have different skill requirements.

Use the skills chart shown in Figure 7 to get an idea of which skills you use most. Just fill in the column for each of the skills to the level that reflects how much you use the skill. The example given in Figure 7 is for someone who works with computers a lot and is fairly dependent on other people at work. Compare your completed chart against Activity 26, where you assessed your current skills. The comparison of the two will help you to identify what you need to work on and lets you plan with your tutor how to develop those skills.

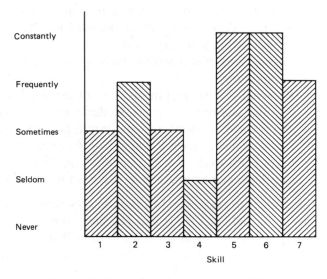

Figure 7   How often do you use the skill?

How are you assessed?

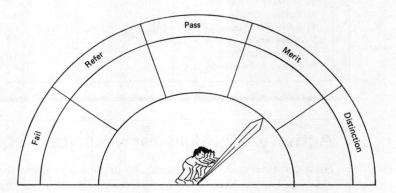

You must pass each element in a module to pass the module but the document you receive from BTEC at the end of your course only includes an overall grade for each module and each common skill area. The grades are on the scale set out below:

- **Distinction** – Exceptional work that is well above the minimum standard in every way and shows a clear and complete understanding of the tasks.
- **Merit** – Work that is well above the basic standard in most areas. It may not be perfect but it is very good.
- **Pass** – Work that meets the basic standard set for the task. No major errors or omissions.
- **Referral** – Work that meets most of the standards for a pass grade but falls down in one or a few areas, so that it can't be passed without some further work. Once this work is done, it will be acceptable for a 'pass'.
- **Fail** – A piece of work that is below acceptable and can't be salvaged by a bit of extra work. It will probably have some major errors or omissions. The assignment or an alternative will have to be done again.

Each piece of assignment work is an opportunity for you to show that you can use the skills and knowledge that you have to get a job done. So, in order to do well on an assignment, you will have to show that you can combine the knowledge and skills you have gained from different areas. You will have to manage the time needed to complete the job and show evidence of initiative and the ability to transfer what you have learned from one situation to another if you are to get a higher grade than a pass for an assignment. The section on final grades later in this unit will tell you more about how these grades are used to decide the final module grades.

When you are given an assignment, you should be told the standard needed for each of these grades. These sets of standards are usually called 'assessment criteria' (see Unit 4). Most colleges will attach the assessment criteria to the assignment, so there will be a sheet with the assignment telling you, at the very minimum, what you have to do to get a pass and a distinction.

Activity 28 Setting the standard

Before you can tell how well you are doing, you have to know how you are being judged. This activity gives you the chance to set the assessment criteria for a task and then to see how well you meet them. The task that you have to do is to build 20 airworthy paper aircraft in one minute. Working in pairs, you should decide how to assess whether someone has done this task well at two levels: distinction or pass. Think about things like setting a minimum distance for the aircraft to fly, looking at the quality of manufacture, the time taken to build them ... .

When you have decided on the assessment criteria, you should join with another pair. Each pair should then carry out the task, in turn, while the other pair assesses them, using the assessment criteria that they have written. When you have finished, discuss with others whether the criteria were right or not. Was it too easy to get a distinction? Or completely impossible to get a pass?

Assessment criteria are intended to give you some guidance as to the standards that would be expected in industry and that the staff are looking for in the assignment. With this information, you can then look at what you'll need to do to get the grades. The assessment criteria won't tell you the answers to the assignment, but they will give you an idea of what the tutor is looking for. They're like the instructions that you would be given at work when you get a job to do – they'll tell you more about what is expected of you, which means that you'll have a better chance of getting it right first time.

If you're going to make full use of the assessment criteria, then you must read them carefully. Here are a number of ways that they can help you:

• They'll tell you what you need to do so that you have an idea of the amount of time that you've got to put into an assignment.

- They let you check, as you're working, that you're on the right track.
- They make it clear where you could have improved when the assignment is returned after grading and you're discussing it with the staff.
- They give you an idea of how the assignment was marked, so that you can plan for the next one. Remember what was said about action plans and reviewing your work in Unit 3.
- They are designed to make sure that you know the standard that will be expected of you in industry.

Weightings

In completing an assignment, you will use many of the common skills, but some of them may be less important and may not be assessed on this assignment. The assessment criteria will tell you which are being assessed, and there will usually be some way of showing which are the most important or heavily weighted in each assignment.

Evidence

For each of the elements in the skill areas and the modules, you will have to show that you can do what is set out before you are able to have it accredited. There are a number of sources of evidence that you can use to show you are capable in a particular element (some of this was mentioned earlier on in the book in the section on APL). For example, in the Business Information Technology module, there is an outcome which states:

Apply a range of business information technology applications

In order to show that you are capable of doing this, you will have to look at the performance criteria to see what you have to do. In this case, they state:

- software correctly set up and used;
- appropriate tasks successfully carried out;
- full use made of features of software.

Now that you know what you have to do, you can start to look for evidence to show that you can do what is required. There are a number of places you can look for this evidence:

- full- and part-time jobs you've had in the past;
- hobbies and interests;
- voluntary work;

- courses you've taken in the past;
- work you've done on the course;
- work experience placements.

Activity 29 Using computers – sources of evidence

Look at the sources of evidence given above and make a note of any experience you've had using computers for each one. When you've done that, you could make a list of the experience you've had based on the most common applications of computers in business – a list is given below and a couple of examples have been filled in.

Word processor: Took CLAIT exam – grade C. Used WORDSTAR. Write letters on home PC using ABILITY for the Squash Club. Used the word processor (SMART) at Johnson's (summer job).

Spreadsheet:

Data base:

Accounts:

Desk top publishing:

When you've decided what evidence you have for a particular element, you'll need to put it into a portfolio – this is really just a collection of the evidence in a folder with a couple of sheets of information to summarise what is there.

So, what is evidence? Any proof that you have done something towards the element you're working on. For such proof to be accepted, though, it will have to be:

- *Recent* – something you did 15 years ago may not be proof of whether you can still do it.
- *Relevant* – having used a very primitive early word processor won't be evidence that you can use a modern package.
- *Checkable* – your tutor will need to be able to check that the evidence is genuine. Just your saying you've done something won't usually be enough. There will need to be a letter, a printout, a certificate or something else that can be checked.

If you aren't sure about some piece of evidence, then ask your tutor. He/she can only say 'no' if it's not acceptable. A lot of the evidence that you'll use will be the work that you do on the course, both in class and in assignments, and there won't be a problem with that – it's recent, relevant and checkable.

When you put a portfolio together, you will need a front sheet like the one shown in Figure 8. With this front sheet, there should be a statement for each of the

Element: Apply a range of Information Technology applications.

Evidence:	PC1	PC2	PC3	PC4	NOTES

Range:					

Element:	Covered by:

Figure 8   Portfolio front sheet

aspects of the range and each performance criteria to show what you have done and how – this guide will help you to look at the areas where you may need to do some more work and also to see what you have actually done.

There is no substitute for you keeping track of this evidence – you are the best person to know what you have done at college, in jobs, as a hobby or in previous courses.

Work experience

Don't forget that you will be assessed on your work experience. It is an opportunity for you to collect all sorts of evidence of what you can do (getting on with people, using computers, working with figures and so on).

Activity 30 If things go wrong

To help you get the most out of your work experience, your tutors will discuss placements with both you and the placement providers to ensure that you are in the right placement. During the work experience, college staff will visit you to check on your progress. You will have assignments to do during your work experience, and these will help you to get the most out of the time spent with the firm.

However, even after all of this, placements sometimes go wrong. There may be a personality clash or the work may not be what you expected. This may affect your assessment because you may not be doing appropriate work or you may not be having the support you need to do your assignment. So, what do you do if things go wrong?

Read the following accounts by two students of how things went wrong. For each case, identify:

1 What has gone wrong.

2 What may have caused things to go wrong.

3 What result the action taken by the student could have on (a) the college and (b) the student.

Richard: My placement was at a large leisure complex, as I was interested in working on the management side of this type of business. I turned up on Monday at 8.45 AM as agreed. I was told to wait until the supervisor came. Eventually, I was informed that the person who was looking after me was off ill, and I was told not to worry and that I would be given some work to do. They then told me to start cleaning the bar. I started to do this, but kept thinking that they would come back and give me something proper to do. By lunch time, I had just about had enough, so I went and told them to stick it!

Madeleine: When I went for my interview, I told them that I wanted to work on the management side of the store. I was told that I would have a programme where I would have various attachments to different managers. I was put on one department where the supervisor was good to me and I learned a lot about how the section was run. The next section I went on was horrible. They had me cleaning shelves all day. I told the supervisor that I wasn't here to do this sort of work, but she told me that all supervisors have to start from the bottom. The next two days I stayed off, but I wasn't sick or anything like that. My tutor phoned me and I told her I was sick. I didn't want to say why I had stayed off. I went back the next day and that same supervisor gave me lots of really boring things to do. I stayed off the next day and the next one too. When I went back, I was called to see the manager, who said that my attendance was poor and that my supervisor had reported that I had been rude to her. I had just about had enough by this time, so I just got up and walked out.

Grading group assignments

Many of your assignments will involve working in a group and producing a piece of work as a group. There are advantages and disadvantages to working in groups, which was discussed in Unit 3. One recurring problem is how to assess individual students as a part of a group that has produced a single piece of work. No tutor would claim to get this perfectly right all of the time. Tutors can never be sure exactly how much work each person in a group has done. Some tutors make an allowance for this by not giving the same grade to each person in the group. So, if Rukshanda, John, Iqbal and Simone are all in a group that has produced a piece of work worth a merit grade, they will get a merit grade overall. If, however, John hasn't done his fair share of the work and Rukshanda has done a lot, then the grades might be:

Rukshanda	Plus one grade	Distinction
John	Minus one grade	Pass
Iqbal		Merit
Simone		Merit

Another way of allowing for the varying contributions of members of a group is to permit some peer group assessment (see later). In such a case, the group members themselves are allowed to vary the grades of other members of the group to reflect the work that they have done, with the help and agreement of the tutor. This is usually a very fair way of allowing for the different amounts of work that people put into pieces of group work, but it requires the members to be honest with themselves and others.

Are there any exams?

You will be no stranger to exams. Throughout your school life, the end of the school year will have meant being locked in hot school halls for two or three hours at a time doing exams. Even with GCSEs, where course work is important, you will have had to do exams. It seems the natural way in which to assess people at the end of their course. But not on BTEC. Not if by 'exam' you mean a traditional two to three hour written exam with no books or talking allowed. There are no exams in BTEC in that sense. What you will find is that you are given final assignments at the end of the year. These final assignments may be in the form of one per module or there may be a number of modules combined into one larger assignment. These assignments will be very similar to the ones that you have got used to doing over the year: they may be a bit longer and they may have some element of time-constrained work in them, but they will not be wildly different from the work that you have been used to.

Activity 31 What no exams?

To help you consider this important question, complete Table 12, ticking one of the boxes in response to each statement. Then compare your answers with others in your class.

TABLE 12 *Your view on exams*

Traditional exams are:	Agree	Disagree
Fair, because they are the same for everyone		
A good test of ability		
A good test of how you cope under pressure		
A good test of how you would perform at work		
Useful because they make you revise		
A good test of memory		
Harder than course work		

If you compare your views with others in your class, you will find differences of opinion. Some people like exams, especially those who are good at them. As a reliable way to test your ability to perform well in business, however, exams are a bit limited. As one manager said:

*No one has ever sat me down in 23 years in business in an empty room
with no phone, no reference books and no one to try ideas on and said:
'Here's a problem, a pen and some paper. Solve the problem and we'll
collect the answer in three hours.' So, why should I trust that situation to
decide if someone will do a job well for me?*

Nevertheless, you will usually find that you will have some special assessment at
the end of your course. Most of the final assignments that you are given will have
a number of features:

- They will test a broad range of skills.
- They will test a broad range of the module content that you have covered.
- They will involve some group work.
- They will involve some research.
- They will give you a chance to show what you can do on your own.
- They will put you under some time pressure.
- They will be set in a realistic business situation.
- They will require you to present answers in written and other forms.

So, although very different from traditional exams, final assignments serve an
important purpose in assessing you across a lot, or perhaps all, of the modules and
skills. In completing the final assignment, you will find that different modules use
different techniques. This is partly due to the differences between modules, and
partly to the differences between courses and between the people who teach them.
So, a final assignment for a module on Business Law might involve you researching
a case, submitting your research as a report, and then arguing the case in a role play
of a court or tribunal hearing; for the International Marketing module, you might
have to do some market research, analyse it and present the findings, with
suggestions in a report as to what the client should do; the core might involve using
a video to tape a presentation to explain to the workforce of a firm about some
changes that are taking place in the business.

   The final assignment will be graded in the same way as your other assignments,
on the scale from distinction through to fail, and will be entered on to your profile
(you'll learn about profiles a bit later on in this unit). Some colleges will treat the
final assignment in the same way as they would treat any other assignment – that
is, not give any extra weight to it – but others will give more weight to it. You need
to find out from your tutor which approach will be adopted.

Who assesses you?

You will have realised by now that assessing students is not an exact science! So,
on the BTEC National there are going to be a number of people and groups

assessing you during the course to give as complete and reliable a picture as possible. You are assessed by:

- college staff
- yourself
- other students
- employers
- BTEC Moderator.

Most of the time, the work that you do will be assessed by the *staff* teams. But *you* have a very important part to play in assessment as well. Although you probably won't be grading your own work very often, you will always have the chance to discuss the grades that you get with the tutors, and you should certainly say if you don't agree with a grade. Your expectations may have been too high but, if that's so, then you need to know why you didn't get the grades that you expected. If there is something outside your control that affected your performance, make sure that you tell your tutor about it.

As well as discussing your grades, you should be reflecting on what you've done and learning from it. This learning from your own work, and not just waiting to be told where you got it right and where you could do better, is part of self-assessment. Unit 3 described how to create your own action plan and how to learn from reviewing the work that you have done.

There are also going to be times when the class that you're in and the groups that you work with will have a chance to assess you and you'll have a chance to assess them. This is called *peer group assessment*. Peer group assessment is used for a number of reasons. It reflects the way that things are at work – the people who you work with have a good idea of what you've done. It also gets you used to considering that other people rely on you to do your share.

Employers will also be involved in assessing you. If you're a full-time student, then you'll be assessed by employers during your work experience on things like the practical use of some office equipment, whether you turn up on time for work, how you get on with other employees and so on.

You'll probably also be given an assignment to do on your work placement and the employer who has given you the placement will help to assess this. If you're a part-time student, your employer will probably be asked to help to assess project work that you do and to comment on work that applies to the firm that you work for.

Both part-time and full-time students may be involved in doing projects for local organisations; these will often involve you in working with employers. You might do surveys for the local council, brief a local business on the different types of photocopier available or help a local small business club to find out about the aid available for exporting. If this is the case, then they will be involved in helping the college staff to assess you.

In all of these cases, it is good for you if the employer gets involved in assessing what you've done, as it ensures that the work that you have to do is real and useful, and that the standards that your work is judged by are those that you'll be judged by at work.

Activity 32 Something's wrong

The following account outlines a student's experience of problems that occurred on a work-based assignment and how they were dealt with. Read it and identify what was wrong with the way the student handled the problems, and suggest a better way to deal with them.

I was asked to carry out an assignment on 6 November at my work. I was given a deadline for the assignment and advised to start early, and not to leave it to the last minute. I told my manager about the assignment and he said that there was no problem.

I started to collect information and made an appointment with my manager – the assignment got off to a good start. Later, I began to hit problems. I could not understand some of the information I had collected and my manager cancelled our appointment, as he was too busy. We arranged another date to meet, but that was cancelled too. By now, the deadline for completion of the assignment was only two weeks away. However, I still thought that there was plenty of time, so I didn't worry.

We started to get very busy at work and I found that I had less and less time to work on the assignment. My manager eventually set aside time to help me and we discussed parts of the assignment – he did help me a lot. He made an appointment for me to see another manager, who would be able to give me some additional help, but she could only see me after the handing-in date.

I stayed off college on the week the assignment was due to be handed in. I hadn't finished it the following week because the other manager gave me a lot of information – she was really helpful. The following week, when I went to college, the tutor asked me for my assignment, but I said I had forgotten it and would bring it in tomorrow. The tutor accepted this but said that if it wasn't in by then I would fail. I thought I would sit up late and finish it. That night we had visitors and I didn't get it finished.

It is inevitable that some work-based assignments will go wrong. The firm may not be able to (or won't) provide the information that you need or your colleagues may refuse you any help. Don't be surprised if some colleagues resent you going to college. If this happens, go and talk it over with a training officer or supervisor, and let the college know that there are problems. Tutors realise that some firms are more helpful than others, and they may be able to contact the firm for you. If nothing can be done, then the tutors can set you a different assignment or make allowances for the problems when grading your work.

A *Moderator* from BTEC visits the centre a few times each year. The Moderator's job is not only to make sure that the course meets the national standards set by BTEC, but also to check and help with assessment. The Moderator will see a sample of students' work to make sure that the marking is of the right standard and that the work is of the same level as work in other colleges. The Moderator will look closely at any 'borderline' work to decide how to deal with special cases, before approving the grades. Moderators usually meet students during their visits to get the students' views on the course, and they will often discuss assessment with the students.

As you can see, many people have a say in the assessment of the work that you do, which means that there are a lot of different views of your ability. Your assessment doesn't depend on the views of just one person or group. This gives a wide range of sources of grades, which means that the final grades that you are awarded should be a good reflection of how you have coped across the course.

Activity 33 Who assesses me?

This activity is intended to get you thinking about all of the different people who have a chance to assess you, and when they are most and least useful.

Table 13 lists all of the people who will assess you. Fill in the advantages of them assessing you and, next to that, the problems. The teaching staff who will assess you have been filled in at the top of the table to give you a start.

TABLE 13   *Details of people who are involved in assessing your work*

Who assesses	What they assess	Advantages	Problems
Staff	Assignments Class work	Know you well	Only see your work at college
Self			
Other students (peer assessment)			
Employers			
Moderator			

As you will see, no one can give a complete picture of you or assess all of your experience, but by working together they can build up a very full and accurate picture.

How are your grades recorded?

On your course, you are going to get a lot of different grades building up to give a picture of you and your work. All of the assignments are graded, often for both a number of skills and for your module performance – there are grades for group tasks, grades from work experience and grades drawn from work in class.

You are going to need to keep track of them all and so are the staff, in order to check your progress and then decide your end-of-year grade in each module. The most common system used is a **profile**. You may already have come across a profile at school or on another college course. A profile is just a special record card. It keeps a record of each of the outcomes in the modules you are taking (you need to pass all of them in order to pass the module) and also keeps track of the assignment grades. Assuming you've passed all of the elements on the module, it is the assignment grades which determine whether you get a pass, merit or distinction overall on a module. Figure 9 shows an example. Most colleges use a profile that allows them to keep track of the skill grades as well as the module grades for each assignment.

Every time that you get a grade, it is entered on your profile. A careful check is kept on your profile by your tutor to see what progress you are making and it is used in determining your final end-of-year grade.

As well as recording your grades, a profile is a useful way to highlight your strengths and weaknesses, allowing you and the staff to look at what you need to build on and what you need to work harder at in order to improve. If, for example, you are very poor at numeracy, this will start to show up quite quickly on the skills part of the profile, as you do badly in the numeracy element of a few assignments in different modules, and the staff may be able to provide some help for you. It is also the most useful tool for your self-assessment and points out the areas that you need to work on in your action plan.

How are your final grades decided?

By the end of your course, a set of assignment grades for each module and for your common skill area will have been recorded on your profile. The staff then have to give you just one final, or end-of-year, grade for each module, plus one grade for each common skill area. These overall grades are the ones that will appear on the certificate that you get from BTEC. These overall grades will not be found by simply averaging the grades that you have obtained in your assignments throughout the year, although these will probably be a pretty good predictor of the final grade that you will get.

Name: Alexandra Hines
Group: BTEC National Diploma in Business and Finance
Stage: Second year

Core Modules:

Outcome	1	2	3	4	5	6	7	8
1	✓	✓	⊘	✓	✓	✓	✓	✓
2	⊘	✓	✓	✓	✓	✓	⊘	✓
3	✓	✓	✓	✓	✓		✓	✓
4		✓	✓	✓		✓	✓	⊘

O = Achieved by APL

Assignment	1	2	3	4	5	6	7	8
1	P		P		D	M		P
2		M	P	R/P		P		M
3	P	M		P	M		M	
4	P			P	D	M	P	P
5	P		M			M	P	P
6	M	P	M	P	D	P	P	P
Final	P	M	P	P	D	P	P	P

Option choices

1 Business Law	1 International Marketing	
3 Business Info Tech	4 Business European Studies	5 Behaviour at Work
6 Accounting	7 Financial Planning	8 Personnel

Modules:

Outcome	1	2	3	4	5	6	7	8
1	✓	✓	✓	✓	✓	✓	✓	✓
2	✓	✓	✓	✓	✓	✓	✓	✓
3	✓	✓	✓	✓	✓	✓	✓	✓
4	✓	✓	✓	✓	✓	✓	✓	✓

Assignment	1	2	3	4	5	6	7	8
1	P	M	P		P	P	M	M
2	M	D		P	P	M		P
3	P		P	M		M	M	
4	P	M	M		M	D		P
5	P	D		M			M	P
6		M	M	P	M	D	M	M
Final	P	M	P	M	M	M	M	P

Tutor comments:

21–11: Two weeks off ill – caught up ok.

13–4: Excellent work placement – has been offered a job.

14–6: Hay fever very bad – extra time agreed for final assignments.

Figure 9   Example of a simple type of profile

The overall grades will also take account of your attendance, the progress that you have made during the year, your contribution to work in class and so on. When you are at work, you often have what is called a performance appraisal, which looks at how well you are doing in your work – your practical work, timekeeping, attendance, attitude and so on. It makes sense, therefore, for such an appraisal to be carried out at college, whether you are preparing for working life, or, if you are working already, for a promotion to a better job or a move to a different job. The process of deciding your final grade for each module is a bit like having an appraisal of your total performance at the end of the year. The same principles apply in deciding both your module grades and your common skills grades.

The final common skills grades, like the final grades for the other modules, are not simply an average of all of the grades that you have achieved in common skills. If, for example, you have demonstrated that you have very good skills in all of the skill areas except numeracy, then you may find that you are referred on this. You may be asked to complete an additional piece of work to bring this skill up to the required standard before you can achieve a pass grade. It may seem very hard not to get your National because you are not up to standard in one common skill area, say numeracy or working with others. However, these skills are necessary in every job and it would be quite wrong for you to get your Certificate or Diploma, indicating to an employer that you have good skills, if you have a major deficiency in one basic skill area.

The final grades are decided at a meeting at the end of the programme by an Examination Board or an Assessment Board, at which all of the staff on the course and the Moderator discuss the overall grades to be awarded to each student, using the profiles and all other relevant information. In the case of a borderline student, the Moderator may ask to see the student or to look at the assignments in order to decide the final grade. Once the Moderator has agreed the final grades, the college will be able to give you your results. This means that you will often be able to collect your results from the college the day after the Examination Board has met, and not have to wait until August. Once the Moderator has agreed the grades, they are sent to BTEC, who will then record them or, if you have finished the programme, issue a Certificate or Diploma to you, although this may not be for several weeks.

Activity 34 Deciding the final grades

Deciding the final grades for a module is not easy. Look at the examples of the grades obtained by some students, given in Figures 10 and 11, and read the information on the students. Then enter what you think the overall grade for each student ought to be. (Assume that the student has met all the necessary outcomes for the module identified.)

Problems – not meeting the standards

During the course, some students have problems: they may, for example, hand work in late, or they may be referred or fail an assignment or a module. In such cases, students will be anxious to know what will happen to their work and their grades. BTEC do not provide many guidelines to help with these problems because they feel that it is for the staff who teach the students to decide, depending on the student's circumstances and the rules that the college has set. Therefore, you should ask the staff about this, since each centre will set its own rules and procedures for dealing with problems like this.

Student: Michelle Jackson

Module: Business Environment

Year: One

Michelle found it hard to settle into college at first, having been in an all-girls school, but she has worked very hard throughout the course and has made good progress since Christmas. She often helps other students if they have trouble with their work in class. She always works well with other students, is very punctual and her class work in general is very good. The report from the bank that she went to for her work placement was excellent, although it said she was a bit quiet.

Grades

Assignments:	1	2	3	4	5	6	Final
	P	P	P	P	M	M	M

Final module grade: ?

Tips
If you average Michelle's grades, then she would probably get an overall pass grade. However, her grades show that she has progressed throughout the course and her report is very good.

Figure 10   Student information for Activity 34

Late assignments

Assignments are about organising yourself to meet deadlines. Obviously, there are some genuine reasons for work being late and, in such cases, your assignment will probably be accepted and graded as if handed in on time. However, if your assignment is late without good reason, your college may have a rule that it cannot

---

Student: Steven Chang

Module: Business Environment

Year: One

Steven worked very hard at the beginning of the course and made good progress during the first term. However, since Christmas his work has not been of the standard that he could achieve. The staff have discussed this with him on several occasions and he has told them that he plays the guitar in a pop group a few nights each week. This means that he has not had as much time to spend on his assignments as he used to have. The group is very important to him. He is coping with his course at college and is fairly punctual. His attendance is a little below average.

Grades

Assignments:	1	2	3	4	5	6	Final
	M	D	P	M	P	P	P

Overall module grade: ?

Tips
If Steven's grades are simply averaged, then he will probably be awarded an overall merit grade. However, his grades show that he has not shown the commitment required and so has not progressed through the course as well as he should have done.

Figure 11   Student information for Activity 34

be marked higher than a pass grade. If you constantly hand work in late, it is possible that it will not be marked at all and the staff may decide to give you a different assignment to do. There are three reasons for penalising you for late assignments:

1 The member of staff may have returned other students' work to them; therefore, it is not fair for you to be allowed to hand your work in after theirs has been returned to them.
2 It is not fair for you to have more time to complete the assignment than other students.
3 It shows that your skills in managing and developing self, and identifying and solving problems may be weak.

If you consistently hand in late assignments, you may pass the particular module concerned but you may be referred or even fail in these two common skills.

Referrals

A referral on an assignment means that you have not quite reached the required standard and will have to do the assignment, or a similar one, again. If you are successful in your second attempt, then, in most colleges, you will only get a pass grade for the assignment, no matter how well you do. However, if you are constantly being referred and then passing on a second attempt, it suggests that you have a weakness in the skill areas, particularly managing and developing self, and this may cause you to fail or be referred on this common skill.

If your final grade on a module is a referral, then you have not covered all of the elements to a satisfactory standard and will have to do some extra work to make good the gaps in what you have done. Your registration with BTEC lasts for five years and can be renewed after that, so it's really up to you to make arrangements with the BTEC centre to prove that you are capable in the areas that you have missed.

The rules on referrals and failures may seem fairly complicated, but the basic ideas are quite simple and are ones that you find in the business world:

• You must reach a minimum acceptable standard; this means that you must pass everything.
• You will usually have a second chance to reach the required standard by doing some extra work.
• The amount of extra work that you have to do depends on whether you are just below the minimum standard, or well below.

Checklist of questions on assessment

This unit should have answered most of your questions. Use the following checklist of questions to see if you know how the assessment will work on your course:

- How are the grades recorded?
- How many assignments are there in each module?
- Are the final assignments based on one module each or do some modules combine?
- Who assesses you? When?
- What form do the assessment criteria take?
- Who is the Moderator? How often does he/she visit?
- What sort of profile is used? What information is on it?
- What factors are taken into account in deciding final grades?
- When are referrals for final grades taken/to be handed in by?
- What are the rules about late work?
- Who is on the Assessment Board?

UNIT 6 Core assignments to practise on

Working under the guidance of your tutors, you will:

- examine detailed examples of the main types of assignments introduced in Unit 4 which you will meet on your course;

- work through the assignments to develop your knowledge and skills.

The first five units have dealt with all of the main elements of your course, leading up to what you are probably most interested in – your assignments and how you are assessed. In this unit, you get the chance to try out what you have learned by tackling some typical assignments. Examples are provided of all of the main types described in Unit 4. These sample assignments are mainly from the core modules, so they can be tackled by all students. Although BTEC doesn't produce standard assignments for centres to follow, the assignments in this unit are a sample of those found on any BTEC National.

Working with your tutor

The activities in the previous units were ones that you could tackle on your own or with fellow students. They were usually fully explained. The nine assignments in this unit, however, are not so complete. You will need advice and help from your tutor to tackle them. The assignments are only given in outline form – there is a basic outline allowing your tutor to modify them if so required. With some slight adjustment, assignments can be used for more than one module of the course or at different times. For example, the assignment 'Late Again!' could be used as it stands fairly early in the course to introduce you to 'Administrative Systems'. With some added background on disciplinary procedures or the law relating to dismissing people, however, it could be used later in the same module, or as a

Personnel Policies and Procedures or Business Law option module assignment. These outline assignments need to be tailored to the course by your tutors. They will brief you on what is expected and tell you the standards expected – the assessment criteria, which can only be set by tutors when they decide where and when in the course the assignment will come.

Assignment layout

Assignments should be clearly set out for your use, just like any business document. The tutors on your course should agree a common layout for assignments so that you can understand every assignment straight away:

- The purpose – what skills and knowledge you will acquire from the assignment.
- The tasks – what you will have to do for the assignment, including telling you which tasks or skills are most important and count most in determining your grade.
- The standards – what standards you are expected to reach for each grade – the assessment criteria.

Not every part of an assignment is equally important. There will usually be some way of telling you which of the tasks, topics or skills are most important in each assignment. Some assignments may test numeracy to a high degree, whereas in other assignments you may not need to deal with figures.

A simple system is needed to give weightings to each task, area or skill covered. A common system used in this book is to put a number of crosses against a skill or task to indicate its importance. For example:

- * the skill is not heavily used in this assignment, but some elements of it are needed in completing the assignment.
- ** the skill is an important part of the assignment.
- *** the skill is a very important part of the assignment.

Colleges have different ways of indicating which are the most important parts of an assignment. Make sure that you note any weightings within an assignment – they are there to help you tackle the assignment successfully.

Sample assignments

For the assignments in this unit and later ones, the layout is as shown in the following example assignment sheet. This is followed by some sample assignments.

Type of Assignment: Whether case study, role play and so on, as listed in Unit 4

Title: A name for easy recognition

Module: Which module it is from

Reasons for choice
Why the tutor has chosen the assignment.

Common skills you use
The main skills you use in this assignment.

Module content you learn
The main topics or areas of the module that you cover in this assignment.

Business setting
The background information and business setting for the problems you have to tackle in this assignment.

Your tasks
Your role in the business setting – what you have to do for this assignment and whether you do it yourself or with others as a group.

Some tips – how to approach the tasks
You should first of all refer to the tips outlined for each type of assignment in Unit 4, but a few extra tips are given for each assignment.

Type of Assignment: Work-based research project (for part-time students)

Title: Company Handbook

Modules: Core

Reasons for choice

This is a valuable assignment for making part-time students aware of the full range of operations in their own company, allowing them to make comparisons with others. It will come early in your course and provides you with lots of information that you can refer back to later in the course.

Business skills you use

Numeracy	**
Managing self	**
Communicating	***
Using technology	**
Design and creativity	**

Module content you learn

Types of organisation
Organisational structures and operations
Giving and exchanging information
Presenting information for a specified recipient
Markets, customers and clients of organisations

Business setting

Your employer wishes to produce a simple company handbook for all employees.

Your tasks

(Individual task, or group task if students work for the same employer)
Produce a handbook for your company employees of not more than eight pages. The main text of the handbook should be stored on disk for updating later. You are free to include whatever you consider relevant, but should cover at least the following areas of the company:

- Origins and development
- Size of workforce and turnover (if available)
- Ownership
- Organisational structure chart
- Main products/services
- Main customers
- Main sources of income and expenditure (if available).

Some tips – how to approach the tasks

(See the general tips on researching information and work-based assignments in Unit 4.) You are very much on your own with this assignment. Plan your time carefully to meet the deadline. Getting the information will probably take longer than you think! The secret of getting the necessary information is to ask for help – from your supervisor or section head, personnel or training manager, or anyone who might have a good overview of the company. Having obtained the information, make sure that your handbook is written clearly, is in your own words and you can answer questions on it.

Type of Assignment: Work-based research project

Title: Business Documents – The Paper Chase!

Modules: Core

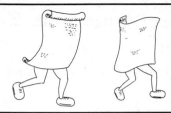

Reasons for choice

Fairly early in your course you need to understand that, for a business to operate efficiently, it must have a set of standard documents that have to be processed in standard ways. This assignment gets you to follow a business activity – ordering goods from a supplier – through all of its stages, noting the business documents used. The assignment is equally suitable for full- or part-time students.

Common skills you use

Numeracy	*
Managing self	*
Design and creativity	***
Using technology	***
Communicating	***

Module content you learn

Business documents
Information flows and systems
Computerisation of records
How people use and misuse procedures and
    documents

Business setting

When ordering goods from an external supplier, any business must go through the following stages:

Internal
requisition   →
of goods

Order placed
with   →
supplier

Goods arrive
at the company   →
from supplier

Goods dispatched
to person   →
requisitioning them

Supplier paid
for goods
received

In addition, the business will record the cost of the original order as well as the eventual payment for budgetary purposes. How these stages are dealt with in any business will vary according to how big it is (there may be a central purchasing and supply department for example), how well organised it is and how many of its operations are computerised.

Your tasks

1 In your company (part-time students) or college (full-time students), take an order and follow it through the various stages outlined above. Collect all of the business documents used during the various stages and discuss how they are processed with the staff who deal with each stage.
2 Produce a company/college handbook for staff use, including all of the necessary documents, explaining how they have to be completed and what happens to them. Also list the main 'do's' and 'dont's' for staff, based on your discussion with those people who process the forms in the company/college.
3 Using a spreadsheet, produce an invoice form that automatically totals the order, adds VAT and produces a grand total.

Some tips – how to approach the tasks

(See the general tips on researching information and work-based assignments in Unit 4.) Part-time students need to get approval and support for this assignment from their supervisor, who may be anxious about the company's documents going to college. With this support, you need to approach the people handling the forms at work, explaining what you need and why. Full-time students will probably be working in small groups and the tutor will point out who to approach. Most people will be willing to talk to you and will be only too pleased to tell you how things go wrong because documents are filled in wrongly (always by someone else!).

Type of Assignment: Case study/role play/presentation

Title: Introducing Fax

Modules: Core, Business Information Technology

Reasons for choice

This assignment involves you investigating the introduction of new technology in smaller organisations. It also requires you to contact equipment suppliers and examine the various claims they make. Your tutor should arrange for a Fax demonstration so that you know what you are investigating.

Common skills you use

Numeracy	***
Managing self	***
Solving problems	***
Design and creativity	***
Using technology	***
Communicating	***
Working with others	***

Module content you learn

Information systems
New technology
Organisational change
Advertising literature
Estimating costs

Business setting

You are the office manager of Japes Textiles, a family business employing 100 staff, which produces and sells a range of high-quality children's clothing throughout the UK. With the European Market, the company intends exporting to Europe. The company has two factories three miles apart. The MD has decided that a Fax system should be installed in the company to speed up communications. Currently, the company uses a local Fax bureau and express postal deliveries. The design and production sides of the company are fully computerised, but the only computerised part of the office is a word processor, which is used by the MD's secretary. The other office staff, six in the main factory and two in the other factory, are very suspicious of new technology.

Your tasks

1 Prepare a report for the MD on the Fax system most suitable for the company. Explain the costs of the system and how it would be organised and managed (group task).
2 Make a presentation of your report to the MD and other managers for production, finance and marketing (group task).
3 Prepare notes for a meeting with the office staff to brief them on the Fax and 'sell' them the idea of its introduction and control over its use (individual task).

Some tips – how to approach the tasks

(See the general tips on case studies, role plays and making presentations in Unit 4.) This is a complex assignment involving many skills and a combination of group and individual tasks. You need to agree on who is to do what and by when first of all, then monitor progress regularly. You need to contact as many suppliers as possible and get their sales literature quickly. A visit to a local Fax bureau or local company using Fax would help you to find out about the range of services offered, and how Fax operations are managed, controlled and costed.

Type of Assignment: Work-based project, questionnaire

Title: Smoking Policy at Work

Modules: Core, Personnel Policies and Procedures

Reasons for choice

This assignment involves part-time students researching an issue in their own place of work so that they can establish guidelines on 'good practice' for all employers. The topic is one that people can get very emotional about, and it can produce conflict between smokers and non-smokers, especially if handled insensitively.

Common skills you use

Numeracy	***
Solving problems	***
Managing self	***
Design and creativity	***
Using technology	***
Communicating	***
Working with others	***

Module content you learn

Health and safety
Consultation procedures
Personal relations and attitudes
Managing conflict
Questionnaire design and use

Business setting

In the UK, smokers now form only one-third of the adult population and, since 1982, they have been a minority in all social-economic groups. The rights of non-smokers to avoid smoke is being increasingly recognised by the majority of smokers. There is cumulating evidence of the potentially harmful effect of 'passive smoking', when non-smokers inhale mainstream smoke (smoke inhaled by smokers and then exhaled into the environment), and the potentially far more harmful 'sidestream' smoke (the smoke from smouldering cigarettes between puffs). Employers have a legal obligation to provide a healthy working environment, and many employers and trade unions are considering the issue of restricting smoking at work on health and safety grounds.

Your tasks

1 Design a questionnaire to discover current practices, the opinions of staff in your company about smoking and whether it should be restricted on your company premises (group task).
2 Survey a representative sample of the workforce in your company (or any other organisation if it is impossible at your workplace) using the questionnaire (individual task).
3 Make a brief presentation of your findings to your class and provide a copy for your employer (individual task).
4 Prepare a code of practice for employers on 'Smoking at Work', explaining the problems, recommending workable principles and practices, and advising on the legal position of employers on this issue (group task).

Some tips – how to approach the tasks

(See the general tips on questionnaires, work-based projects and giving presentations in Unit 4.) Your tasks are clearly set out for you in this assignment. Your main difficulty will be getting up-to-date information on this topic, but the Health Education Officer at your main hospital would be a useful local source. To carry out your survey, you need the approval of your employer, and your safety officer and trade union officials may be willing to advise. You will meet some suspicion, and perhaps hostility, in carrying out your survey. Explain that it is a college project and that you are concerned solely with health and safety aspects of smoking, not with whether smoking is or is not an acceptable habit.

Type of Assignment: Role play

Title: Sex Discrimination

Modules: Core, Business Law

Reasons for choice

This assignment involves developing a detailed understanding of the law on discrimination in employment and assessing your own interpersonal skills. You have to act a part in a role play, according to the rules and procedures of a tribunal. The role play may take up a full day, or possibly more, in addition to preparation time.

Common skills you use

Solving problems	***
Managing self	**
Using technology	***
Communicating	***
Working with others	***

Module content you learn

Rights and responsibilities of employers
   and employees
Conflict resolution
Industrial tribunals

Business setting

Mrs Judith White has taken her employers to an industrial tribunal on the grounds of sex discrimination when she applied for a recent promotion to section head. She had more experience and better qualifications than other applicants, who were both males. They were both interviewed and one was promoted. She was told by her boss that the section head needed to be a man 'because people in the section wouldn't stand for a woman'. He also said that the maternity leave she took two years ago when the firm was busy with orders wasn't the example expected from a section head.

Your tasks

A number of roles are allocated for the tribunal hearing with a brief provided for each role. Depending on the size of the class, two or more tribunal hearings could be held, so everyone plays a role. The main roles are:

- Mrs White and her legal representative
- The manager of the company and his legal representative
- Tribunal members.

Each student must:

1 Play a role in the tribunal hearings.
2 Write a letter from the tribunal to Mrs White and the company, giving the reasoned judgement of the tribunal.
3 Write an article for the local evening paper, reporting the case.

Some tips – how to approach the tasks

(See the general tips on role plays and researching information in Unit 4.) You need to do your research thoroughly before the tribunal, or you will quickly get caught out. Make sure that you know your role and the line you intend to follow. Don't get ruffled and try to keep calm, whatever the provocation. Remember, you are being assessed on how you act your part as well as how much you know about the law on sex discrimination.

Type of Assignment: Role play

Title: Late Again!

Modules: Core, Business Law, Personnel Policies and Procedures

Reasons for choice

This assignment is set primarily to develop your skills in dealing with people at work and trying to see other people's point of view. It also emphasises the fact that the standards expected of employees by their employers, in this case concerning punctuality, may be much higher than those tolerated at school or college.

Common skills you use

Solving problems	***
Communicating	***
Working with others	***
Managing self	*

Module content you learn

Human relations
Conflict resolution
Obligations of employers and employees
Disciplinary procedures and counselling

Business setting

Tina is the office supervisor in a large firm of solicitors. Maisie is the receptionist/telephonist, who starts work at 8.45 AM and finishes at 4.30 PM. Maisie has been a bit late three times since she joined the partnership a month ago, although she is good at her job in all other respects. She offered no explanation for the lateness but has apologised profusely. Today, Maisie arrived at 9.10 AM, well after the office was open for clients. The senior partner has told Tina to sort out the problem, even if it means dismissing Maisie. He also criticises Tina for not sorting out the problem the first time Maisie was late.

Your tasks

Tina calls Maisie into her office at 9.30 AM to sort out the problem. You are to play the part of one of them using the following brief to guide you in the meeting:

1 Tina's brief – you are over 50 and have been with the solicitors for 30 years, working your way up from office junior. You believe in high standards and expect this of all of your staff. You don't want to lose Maisie, who is a good worker, but you won't tolerate sloppiness. You also feel her actions have damaged your standing with the partners.
2 Maisie's brief – you are 24, black and divorced with a two-year-old son, Paul. You need the job badly and enjoy the work. You feel Tina is a bit old fashioned in her ways and she sometimes criticises your dresses as too brash. You dare not tell her that the reason for your lateness was that you couldn't get Paul to the childminder's in time for work.

Some tips – how to approach the tasks

(See the general tips on role plays in Unit 4.) The key to this assignment is to think of yourself as Tina/Maisie – to think and behave as they would during the meeting, rather than as you would do yourself. If you are male, this may feel strange, but remember that one of the purposes of the assignment is to get you to see things from other people's point of view! You should prepare carefully for the interview. The interview may become tense, but both sides have a responsibility to maintain a professional atmosphere. Listen carefully to the questions put and use body language, as well as words, to present your arguments. Both parties need to be sure of their rights with regard to dismissal. Remember that, after the interview, Tina and Maisie will probably still need to work together in co-operation.

Type of Assignment: Case study, research project

Title: The Pensions Market

Modules: Core

Reasons for choice

In this assignment, you will work with other students to research and present findings relating to the growing financial services sector. You will see how businesses must research customer needs and market their services against competitors. This assignment could come fairly early in your course as it is a good way of introducing you to group working.

Common skills you use

Numeracy	***
Solving problems	**
Managing self	***
Design and creativity	***
Using technology	***
Communicating	***
Working with others	***

Module content you learn

Personal financial planning
Financial services
Interpreting financial information
Marketing and selling
Profits and investments
Competition
Legal regulations
Population trends

Business setting

AJP Assurance plc is a respected and well-established company with agents across the country. The company has concentrated heavily on traditional insurance but is now looking at the boom in pensions as a major market opportunity for the next 20 years. You work in one of the company's research and intelligence units, which has been asked to prepare a study of opportunities for the company in this market, and a marketing plan in conjunction with the marketing unit.

Your tasks

(Group tasks)
1 Prepare your study for the next Board Meeting, indicating market potential, the main services provided by other companies, and any trends or opportunities that AJP might take advantage of.
2 Make a presentation of your study.
3 Prepare an outline advertising campaign for TV to promote the company in the pensions market.
4 Prepare a 30-second TV advert on video.

Some tips – how to approach the tasks

(See the general tips on case studies and researching information in Unit 4.) This is a fairly large project and your group needs to agree quickly on who is to do what and when. Regular checks on progress are essential. You won't find it difficult to get information on state pensions, and local banks and insurance companies should willingly provide you with information on private pension schemes and investments. Remember that the size of the market is heavily dependent on population trends, especially the number of people approaching retirement, which can be easily discovered from government statistics in your college or local library.

Type of Assignment: Research project for local employer

Title: Survey of Customer Satisfaction

Modules: Core

Reasons for choice

This assignment has been chosen because it gives you the chance to work on a real problem supplied by a local organisation. The research report and presentation have to be organised to a timetable and to a standard that is acceptable to the outside organisation. In effect, you are acting as a consultant, helping to solve a real problem, and tutors are always on the look out for such 'commissions' for students.

Common skills you use

Numeracy	***
Managing self	**
Solving problems	***
Using technology	***
Communicating	***
Working with others	***

Module content you learn

Marketing
Questionnaires
Customer requirements and satisfaction

Business setting

Your local bus company has been heavily criticised for poor services in local newspapers and by local councillors. The company has good links with the college and has approached your department to carry out a survey of passengers, over a one-week period, to discover their satisfaction with bus services and the main improvements they would like to see in services. The company has a list of the most common complaints. You will be allowed to travel on the buses to carry out your survey and to question passengers in the main bus station.

Your tasks

This is a whole-class exercise which the tutor will want to maintain close control over. You will be provided with a clear brief, indicating the main tasks as:

1 Determining the sample to be surveyed.
2 Designing, producing and administering a questionnaire for the sample.
3 Analysing the results using a computer and producing a report for the company to acceptable business standards.

Some tips – how to approach the tasks

(See the general tips on researching information, questionnaires and presenting your answers in writing in Unit 4.) This assignment requires the tutor, the class and the company to work closely together. Each of the parties needs to know what is expected of them and by when. Remember that, to carry out the work, you will be operating outside of the college with only limited supervision, and you may need to work outside of normal college hours to get a good sample of passengers. Your assessment will probably involve the company as well as your tutor, and in such a group project some peer group assessment is desirable.

Type of Assignment: Organising an event

Title: A Promotional Event

Modules: Core

Reasons for choice

In this assignment, you will work with other students to provide information on your course to employers and parents, to give a favourable impression of the course and its students. You will work together on a real event for members of the public. It involves a high degree of initiative and responsibility, and will usually come towards the end of the course.

Common skills you use

Solving problems	***
Managing self	*
Design and creativity	**
Communicating	***
Working with others	***
Using technology	*

Module content you learn

Giving, exchanging and disseminating information
How to work as a team
Organising to meet a deadline
Communicating with individuals and organisations
   outside of college
Budgeting

Business setting

A promotional evening is to be arranged for parents and employers of students on your course. The event is to be organised outside of college at a suitable venue. The purpose is to promote the full- and part-time courses by displaying student work and providing promotional material on BTEC. It is also hoped that the event will encourage employers to provide more support for the course through visits, project or work experience for students. The evening should be informal with light refreshments provided. All reasonable costs will be met by the college.

Your tasks

(Group tasks)
Planning, organising and conducting the evening are your responsibility, but tutors will allocate some tasks and monitor progress. Your plans should include:

- Booking a suitable venue and refreshments.
- Organising suitable displays and promotional activities.
- Inviting guests.
- Producing a programme for the evening.
- Receiving guests.
- Securing maximum publicity for the event.

Some tips – how to approach the tasks

(See the general tips on organising events in Unit 4.) This is probably the most complicated assignment that you will do on your course. You need to plan backwards from the date of the event to ensure that all of the jobs are done, and in the right order. Write out a complete list of jobs and who is to do what by when. Make regular checks on how things are progressing. Make sure that you have a 'dry run' of the event in the venue beforehand. Pay attention to detail (spelling mistakes in the invitations don't inspire confidence!). Your assessment will probably include some element of peer group assessment, so keep an eye on the contributions of various team members.

UNIT 7 | Assignment programmes

Working under the guidance of your tutors, you will:

- examine the ways in which individual assignments are combined to form a programme;
- improve your learning and studying skills;
- work through a sample of assignments from different stages of your course.

This unit will also provide ideas to tutors to help them plan their assignment programmes.

What is an assignment programme?

You are now thoroughly familiar with the way in which you learn and are assessed on your course through practical assignments. You may have tackled many different types of assignment by now, as outlined in previous units. On your course, the tutors will have worked out an assignment programme – that is, a plan of assignments. Often, you will be told about this full programme at the start of the course.

How many assignments are there in an assignment programme?

This is the most obvious question that you will want to ask your tutors. Unfortunately, it's a bit like asking 'how long is a piece of string?' BTEC requires that all of the core and option modules are fully assessed, but does not stipulate how many assignments are needed to do this. This decision is left to your tutors with the supervision of your BTEC Moderator. When the course is highly integrated, students may only have to do a few, large assignments, to cover the core and option modules. On a less integrated course, students may do more smaller assignments. In the end, though, students should do the same amount of work to the same standards on all BTEC National Business and Finance courses.

Some basic rules

Although there is no ruling from BTEC on how many assignments you should do, there are some basic rules that tell you what you can expect, and in some cases demand, from each tutor.

For each module or set of modules, you should have an assignment programme that is:

- *Planned* – the number, type and timing of assignments is organised in advance.
- *Varied* – includes case studies, role plays, projects and other types of assignment, as described in earlier units. You should not just have one type of assignment to do.
- *Progressive* – starts with easier assignments, so that you can settle in, then moves on to harder ones; and often finishes with a final assignment to bring it all together.
- *A mixture of individual and group assignments* – doing only individual or only group assignments is not usually a good way to assess your abilities.
- *A mixture of assignments to be done in college and outside of college* – in businesses where you work, do work experience or simply visit. Your assignments should take you out into the business world and the community.
- *Comprehensive* – covers the module content and the common skills. All modules, including options, should develop and assess skills; therefore, all your assignments for all of the modules should include skills.
- *Well produced and documented* – the assignments you get should be clear, well-produced business documents, properly set out and clearly produced.

Managing self – your learning and studying skills

When your tutors plan and organise your assignments properly, they are organising your course in a proper businesslike manner. They expect you to learn and study in the same businesslike way. You will get a lot of assignments, some of which will have to be finished around the same time. You are expected to:

- Plan and use your time, especially any free time, for doing assignments. No one will be standing looking over your shoulder!
- Hit the deadlines set for assignments – or suffer the consequences!
- Ask for help and guidance from tutors if you need it.
- Produce assignments to a proper business standard – word process them if you can.

- Review how you have performed when your assignments are graded, think about what you have learned and institute an action plan to improve next time.

Getting through the assignment programme is a challenge. Most students can tackle their assignment programmes successfully, if they plan properly. It is your chance to show that you have the most basic of the common skills listed by BTEC: managing and developing yourself effectively. If you do not develop these skills and recognise their importance, you will have problems on your course and in employment. The following activity helps you to think about how to improve your organisation for assignments.

Activity 35 Deadlines

Every student has problems in hitting assignment deadlines from time to time. Complete Table 14 by listing the reasons why this is so, the consequences of not hitting deadlines and things that would help to hit deadlines. Some pointers have been entered in the table to get you started.

TABLE 14   *Meeting deadlines*

I miss (or nearly miss) deadlines because:	If I miss a deadline:	I can hit deadlines better if I:
1 The tutor hands it out late	1 I get a lower grade	1 Keep a diary of assignments
2 I leave it to the last minute	2 I'll be late for other assignments too	2 Get started early
3 It clashes with another assignment deadline	3	3
4	4	4
5	5	5
6	6	6

Now discuss the points you have included with other students in your class and select those tips that are the best for you.

Sample assignment programmes

Although assignment programmes will vary, all of them have a number of common features, as illustrated in the following examples. As far as possible, the examples are laid out in the same way as those in Unit 6. There are examples of:

- Induction assignments – something you can tackle at the start (induction) of your course, which does not need a lot of knowledge of business but gets you thinking in the BTEC way.
- Work experience assignments – ones that help you to prepare for and learn from your work experience.
- Multi-stage integrative assignments – ones that have several stages running over several weeks or even months and are designed to pull together all of the modules and skills.

Many of the sample assignments in Unit 6 could be combined to make assignment programmes but here are a few more.

Type of Assignment: Induction assignment

Title: Personal Finance

Module: Part of general introduction

Reasons for choice

This assignment is part of a number of 'ice-breaking' exercises, which are arranged at the start of your course to get you used to assignments and working with others. It deals in a simplified way with basic business ideas of income and expenditure and the financial year.

Common skills you use

Numeracy	***
Managing self	**
Using technology	**
Communicating	***
Working with others	**

Module content you learn

Income and expenditure
Personal budgeting

Business setting

Your personal budget is similar to that of a business. You have income from various sources and expenditure on a range of items. Some items of income and expenditure are easily controlled while others are less so. In addition, you may receive benefits in kind, rather than cash, from your family, employer or other sources.

Your tasks

1 Complete your personal income and expenditure table below for the current financial year.
2 If inflation increases on average by 9% over the next financial year and your income increases by only 5%, how could you manage your budget?
3 Compare your work with another student, noting the main differences and reasons for them.

Cash income		Cash expenditure		Benefits received in kind	
Source	Amount	Item	Amount	Item	Amount (Est.)

Some tips – how to approach the tasks

Remember to include all of your income and expenditure; do not be shy about discussing your financial matters with fellow students and staff. This is part of the ice-breaking exercise.

Type of Assignment: Work experience

Title: Getting the Most out of Work Experience

Modules: Core

Reasons for choice

To get the most out of your work experience, you must prepare, record what happens and look back afterwards to see what you have gained. Your tutors will have some record systems, possibly similar to the examples given here.

Common skills you use

All skills may be used depending on the job.

Module content you learn

A wide range depending on the job.

Your tasks

Stage 1 – preparation

You are preparing for your first work experience, so you need to know what is expected of you in your placement.

1 On the table provided, list the differences you expect to find between being a full-time student in college and being in work. Some clues are given to get you started.

Clues	College	In work	Your work placement
Hours Discipline Safety Absences Dress			

2 Make an appointment to visit your placement prior to the official work experience period. Then complete the right-hand column of the table so you know exactly what to expect.

Stage 2 – recording what you did on placement

When you are on placement, you need to keep a daily record of what you did so that you can look back at the placement afterwards and see what you learned.

3 At the end of each day, complete a daily log like the following as fully as possible, with examples and notes that you can refer back to later.

Day	Worked with people	Worked with equipment	Worked with business documents	Met customers
Morning Afternoon				

Get your work supervisor to sign your log each day, if possible.

Stage 3 – deciding what you learned after placement

You have finished your placement and need to look back immediately to see what you learned from it.

4 Refresh your memory on what you did from your daily log. Then complete the following self-assessment record by ticking the appropriate column. Your placement supervisor will probably also have completed an assessment record, which you should ask your tutor to show you.

Reaction	Yes	No
The work was varied		
The work was interesting		
The work was hard		
The work was easy		
The work was boring		
The work was tiring		
My supervisor was helpful		
My supervisor was friendly		
Other staff were helpful		
Other staff were friendly		
The visit by my tutor was useful		
The placement was useful for my course		

5 After refreshing your memory, write down the three most important things that you learned.

The three most important things I learned were:
1
2
3

6 Complete the following skills self-improvement table, based on what you did and how you thought you improved. Be honest with yourself!

Skills	Things I did on placement that involved using skills	How I improved my skills during placement	Skills the placement showed I need to work harder at
Numeracy			
Solving problems			
Managing self			
Design and creativity			
Using technology			
Communicating			
Working with others			

Some tips – how to approach the tasks

Your period of work experience is the closest that you will get to learning through doing until you actually get a permanent job after your course. The lessons you learn on placement – whether it's a good or bad placement – are quickly forgotten, and sometimes you will not realise the significance of something that happened on placement. So, preparing for placement, recording while on placement and putting on paper your impressions immediately afterwards are all part of getting the most out of work experience. Even when things go badly on placement, you can still learn from this. Remember that learning through doing requires you to reflect on what you have learned and to pick out the general points from your particular experience. This is why you will be expected to record what happened and how you felt and then to discuss these points back at college.

Type of Assignment: Multi-stage integrative assignment

(lasting over several months)

Title: Mini-business
Modules: All

Reasons for choice

One of the best ways to learn about business, if you are a full-time student, is to run one for a while. In this assignment, you will work as part of a group to set up, run and then close down your own small business. You will learn a lot about what it takes to survive in business, because you will have to make your own decisions, although in this case there is little risk to you if things go wrong. You will also learn how important team work is for a successful business. Your tutor will keep a check on your performance.

Common skills you use

All skills may be used.

Module content you learn

Business growth, development and closure
People and teams in business

Business setting

As a part of your course (possibly in the second year), you are required to set up and run a small business. The college will provide you with all of the resources and help that it can, especially advice. A local bank will make a bank loan available to you. The business can take any form that you like, as long as it is legal, decent, honest and truthful. As part of the assessment for this exercise, you will be required to keep a 'portfolio' of all of the work that you have done in setting up and running the business, and this will need to be submitted for grading. It is in your interest to keep this up to date and well organised. In addition, you will have to provide a written report on what you have learned from the exercise and a grading of the contribution of each team member, including yourself, to the business.

Your tasks

The assignment involves a number of stages and tasks over the year, following roughly this timetable:

Stage 1 (Sept.–Oct.)   Deciding your business
Stage 2 (Oct.–Nov.)   Researching the market
Stage 3 (Nov.–Dec.)   Setting up your business
Stage 4 (Dec.–March)   Running your business
Stage 5 (April–May)   Closing your business

The exact tasks you need to carry out will vary according to your business, but you will certainly need to work through the following, although not necessarily in the exact sequence they are listed. The list is to ensure that you are systematic, although several tasks can be started at the same time.

Stage 1 – deciding your business

1 In small groups, identify those business ideas that you could try out which do not require much capital.
2 Select an idea that interests your group and seems commercially feasible.
3 Identify the skills, knowledge and resources you will need to run your business.
4 Identify the skills, knowledge and resources your group already has and those it needs to develop to run the business successfully.

Stage 2 – researching the market

You now have a clear idea of what your business might be but it needs to be tested, to see if there is a market for your goods/services.

5 Design a questionnaire and carry out a survey to find out what your potential customers think of the goods/services you intend to market.
6 Carry out any other research (of your competitors for example) to assess the market for your business.
7 Identify your capital requirements to start up and run your business, and calculate costs, sales and profits for the business over the period of the assignment. Draw up a break-even chart.

8 Decide whether your business is going to be viable. If not, go back and work on your idea, or a new idea, until you get a viable business idea that stands the market 'test'.
9 Draw up a business plan for setting up and running your business for the rest of the assignment and present it to a local bank manager at an interview arranged by your tutor. You should include:
   • The business idea
   • Market demand
   • Resources required
   • Organisation and management responsibilities
   • Costs
   • Projected profits
   • Any other relevant details.
   The bank manager will decide if your business plan merits a bank loan and will offer advice on how to run your business successfully.

Stage 3 – setting up your business

10 Decide who is to do what and when in the business, and issue written contracts to each group member stating his/her duties. Establish a log book to record the work done by each member.
11 Decide how to advertise your business, design suitable promotional materials and agree a marketing plan for the full period of the assignment.
12 Open a bank account to call on the loan agreed by the bank manager.
13 Set up a computerised bookkeeping system to record all of the takings and expenditure, and make arrangements for handling and storing cash.
14 Set up a computerised stock control system to ensure that you keep track of stock and can place orders in good time to avoid 'stockouts'.
15 Place your orders with suppliers, and make arrangements for receiving and storing goods.

Stage 4 – running your business

All of your preparations have been made and your systems are ready. Now go for it and make your first million!

16 Carry out your marketing plan, adjusting it over the year according to your success and market opportunities.
17 Make and sell your goods/services, keeping careful records so that you can respond to market opportunities, introducing new lines and discontinuing existing ones, if necessary.
18 Keep your financial and stock control records up to date and also your log book of work done by team members.
19 Hold regular formal meetings of the full business team to monitor progress and take decisions promptly, as required. Ensure that there is an agenda produced and the minutes are taken.

Stage 5 – closing your business

The financial year and also your course are coming to an end. It is time to stop trading.

20 Cease trading and complete your end-of-year accounts up to 31 March.
21 Dispose of unused stock.
22 Distribute profits after repaying all loans.
23 Present your portfolio of the work you have done in setting up, running and closing the business.
24 Write a report on the lessons that you have learned from the exercise, what you think you have gained and what areas the exercise shows you still need to work hard on.
25 Give each team member a grade (distinction, merit, pass and so on) for his/her overall contribution to the assignment.

Some tips – how to approach the tasks

Do not make the mistake of thinking that the purpose of the exercise is simply to make a profit and that, if you make a big profit, you are sure to get a good grade. The purpose is for you to learn about all of the stages in the life of a business, in addition to the organisation and paperwork that go with them. You will be assessed according to how well you handle the various stages and set up procedures in a proper businesslike way. You will also be assessed on how you work in a team and make thoughtful, imaginative and well-reasoned decisions. How you handle any awkward or lazy members of your team will be noted.

For such a lengthy multi-stage assignment, you need a clear plan and timetable for the full period of the assignment, although these need to be flexible in case things do not turn out as you expect. The 25 tasks provide you with a sort of checklist to help your team agree on an initial timetable, although you do not have to follow them in the exact order suggested. You can work on several tasks at once, but generally you need to complete one stage before trying to do too much work on the next.

As the assignment progresses, do not be afraid to be flexible over your plans or timetable. Take advantage of new opportunities, such as a new location to sell your goods from or a new product line you might come across. And do not just wait for things to happen if sales go badly – look for new customers, or hold a sale or promotional event.

Finally, this assignment lasts a long time, so you will find it useful to keep a personal diary of what happened, why and who contributed what, in addition to the business log book. Make a special note of any key points that you learn. This personal diary will be invaluable in helping you to complete the last three tasks, which are an important part of the assessment.

UNIT 8 Sources of help and information

In this unit, you will:

- find out the *sources of help and information*;

- be given an idea of *how to use the help that is available to you.*

- Where on earth am I going to find out about unemployment in London?
- Who can tell me what the FTSE index is?
- What does the average family spend on food each week?
- What's the share price for British Telecom today?
- How can I find out how an industrial tribunal works?
- What sort of job would suit me after I finish my course?

These are the sorts of questions that you might have to ask at some point in your course, but there are a lot of other things that you may need to know as well. You'll be told about a lot of them in your classes, whereas for others you're going to have to find out for yourself. This unit aims to introduce you to some of the main ways in which you can get help and information, and to give you an idea of what they offer and how to get the best out of them. The unit is broken up into the following sections:

- Help in your college.
- Self-help.
- Help locally.
- Contacting people for help.
- Help further afield.

Help in your centre

There will be a number of different sources of help and information in your college that you can use. Exactly what is available will depend on the college or centre where you're studying, although most provide a certain basic set of services for students. They may use different names – the library may be called the learning resources centre – but they'll all do most of the same things, although some are better equipped than others.

Activity 36   Sources of information in your centre

For your own centre, make a list of the places where you can get information to help you with your assignments. Next to each one, note the information that you can get there and any other useful details, such as opening times and location. Some suggestions have been added to Table 15 to get you started.

TABLE 15   *Sources of information in your centre*

Facility	Information	Notes
Library	Statistics Textbooks Prestel	Open 9 AM to 9 PM Monday to Friday Ground floor, main block

---

The main facilities that you will find will probably be:

- the library
- a business resource room.

Depending on the centre, there may be others that are also open to you. Let's have a look at what is available from each of these facilities.

Library

Libraries in colleges have a lot to offer you. Many of them have areas where you can study and discuss things, and they provide a lot more than just books. Figure 12 shows a plan of a college library; not all of the things on the plan would be in every library, but equally some may have a lot more.

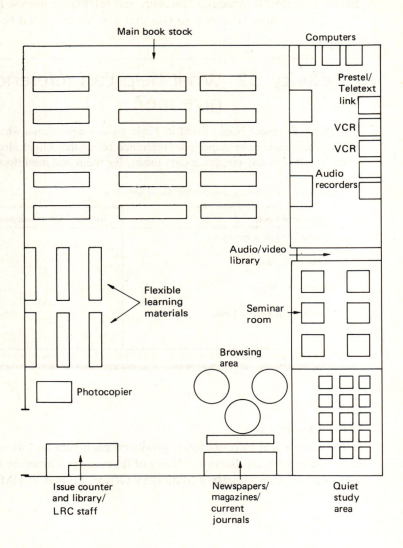

Figure 12   Plan of a library showing the main facilities

Books

Reference books

You may already know about many of the reference books in the library, such as dictionaries, encyclopaedias, thesauri and telephone books. However, there may be others that may be useful to you that you haven't used before.

Activity 37 What help can reference materials give me?

Find the reference books listed in Table 16 and write down what they tell you. Then make a note of some other reference books that might also be useful. If you have access to any computer data banks, list them and the information contained.

TABLE 16 *Reference books that could help you*

Reference book	Information contained
Croner's Employment Law	
Whitaker's Almanac	
Who Owns Whom	
Commercial Yellow Pages	

Official statistics

A lot of statistical information is produced each year and most libraries will stock some of the main publications. Many of them are produced by the Government and published by Her Majesty's Stationery Office, known as HMSO.

Activity 38 Using official statistics

Table 17 lists some of the main statistics collections. Find them in your library and fill in what information they contain.

TABLE 17   *Sources of statistics*

Publication	Information contained
Family Expenditure Survey	
National Income Blue Book	
Social Trends	

There are other statistics and many of them will use the same basic figures in different ways. You need to look at the figures carefully and remember the old saying 'There are lies, damned lies and statistics'.

Textbooks

You will need to use many texts across your course. The ones that you need will be recommended to you by your tutors; however, if you want to read ahead or you need a book to help to get something clearer in your mind, your tutors will be happy to recommend one to you. You will probably have to fill in a form before you can take any books out of the library. You may also need to have it signed by your tutor.

One thing to watch out for, especially in areas where the content becomes out of date quickly, is to make sure that the books are fairly modern. A book on licensing law, for instance, will be out of date if it's more than three years old. Most libraries put popular books on 'short loan', so that they can only be taken out for a short time – this is to stop people holding on to books that other people need. If a popular book isn't on short loan, ask the lecturer for the module to get the library to put it on short loan. Other books may be 'reference only', so that they can't be taken out at all, maybe because they are going to be needed by a lot of students at once. If there is a small amount of information that you need from a book, you may be able to photocopy it on a copier in the library, but ask the library staff about this, as you could be breaking copyright by copying.

Periodicals and newspapers

Most libraries take a selection of newspapers. They are a good source of up-to-date general information and you should try to read one each day. There is no better way to keep up to date with developments in the business world. It is important that you read one of the quality papers, as these have daily specialist sections on business as well as reporting major business news stories. You will find the *Financial Times* the most comprehensive but the *Guardian*, *The Times*, the *Independent* and the *Daily Telegraph* also give excellent coverage. Beware of the biases of the different papers. They'll also give you information like share prices, currency rates, interest rates and business news about specific companies. When there is something major happening – like the Budget – they will have special articles, and these are normally up to date and easy to read; in fact, they are ideal for background reading to an assignment (but if you use information from a newspaper, make sure that you acknowledge where it has come from). It is useful to keep cuttings or copies from useful articles, so that you can file them with your course notes.

Journals and magazines

There is a whole series of magazines that could be useful to you, depending on the modules that you take. Here is a list of some of the main ones:

- *The Economist* – a weekly magazine dealing with political matters as well as economics. It concentrates on how the economy affects business. This is a 'must' for all business students to read.
- *Business Law Review* – a monthly review of changes in the law that are going to affect businesses with some good articles on topics like employment law.
- *Marketing Week* – a weekly review of marketing and the main developments in the industry.
- *Time* – a weekly review of world events. This is an American journal so you will get a slightly different view from our papers.
- *Campaign* – a weekly journal for the advertising industry. It provides some very good case histories of campaigns that worked and some that didn't.

- *Industrial Relations Journal* – a monthly round-up of the latest issues in industrial relations and the main court decisions that have taken place.
- *Personnel Officer* – a weekly, covering the personnel industry and training. It has some very good articles.
- *Economics* – a monthly magazine aimed at economics students. It can be useful for clear explanations of some of the theories behind what you are covering.
- *Spectator* – a weekly magazine that is based on topical issues.
- *New Statesman and Society* – a weekly magazine dealing with social and political issues. You only need a little bit of thought to see how some of what they cover will affect business.
- *Labour Research* – a monthly journal that covers business and politics from a Labour and trade union viewpoint. It contains well-researched articles.

If a journal is particularly useful, you'll be told about it by the tutor for the module. But spending a free hour or two looking through some of these, without being told to do so, is a good habit to develop and a useful way to use some free time. You may also find something that will be useful in an assignment at some stage.

Prestel/Teletext

These information services allow you to get information on screen from national data bases on all sorts of subjects, from share prices to holidays. You'll probably have to be shown how to use the system first and there may be some constraints on using it, as it can be quite costly to run.

Video viewers

These allow you to view the collection of video tapes on educational topics that libraries keep. Normally, there will be only a few tapes taken from the TV, as the college will have to pay a special fee for the programmes that they tape live. You can also use them to view useful TV programmes that you or someone else has taped.

Computers

Many colleges now provide computers in the library for students to use. Arrangements vary for their use: some colleges ask students to sign for a key to the computers and software and to return it when they've finished; others leave the machines on 'open access', so that anyone who has permission can use them. Almost certainly there will be some control on who uses them and when, so ask before trying them.

Tape players

These provide students with a facility to listen to some of the educational material that is held on tape in the library, from language learning to management tips.

Seminar rooms

Many libraries have seminar rooms that staff can book to bring students to when they want to do some library-based work and discuss the work or the information with the students as they find it. Some of these rooms can also be used freely by students, when not booked by staff, and this provides a place where a group can come to, to do some work and talk without disturbing other people.

Business resource rooms

Many business departments also have their own resource rooms that their students can use on an 'open access basis'. The facilities typically found in one of these rooms include the basic office equipment that is needed to prepare work in a professional way, as well as computers, typewriters and sources of useful information. Many such rooms have someone in charge, who can help you with software that you don't know well, issue materials or sell things like files and pens, if you need them. Departments may also have a training office that takes in work and operates like a commercial office for students to practise in – they may be able to do typing or copying for you at reasonable prices.

Self-help

There are quite a few ways in which you can help yourself and other students, by making use of the resources that you have as a group. Here are a few ideas:

- Make sure that you look out for useful programmes on the TV. There are a lot of programmes that deal with topics that will be useful to you, such as *The Business Programme, Panorama, World in Action* and other documentaries. If a programme is going to be useful and you can't see it, try to get someone to tape it, so you can watch it later.
- The 'news' is well worth watching each day. If you can try to catch one of the main news programmes, especially Channel Four News, this will give you quite a lot of useful background information, as well as dealing with important business topics, like the Budget, in some depth.

- There are quite a lot of radio programmes that you can get useful information from. If you aren't sure if something is going to be useful, try to take it in turns to listen to, and tape the programme if it is useful. Regular programmes to watch out for are *PM* (Radio 4 – news review), *Today* (early morning news review and magazine programme) and *Yesterday in Parliament*. There will also be some documentaries or short series programmes that are well worth listening to at different times across the year. Try to look at the radio schedules regularly to see if there are any useful programmes that you could listen to.
- As a group, you can arrange for each person to keep an eye on one or more magazines regularly and to let the rest of the group know if there is something useful that they ought to read.
- Keep a cuttings file for topics that you know you are going to look at; for example, if there is an article on the European Market and you know that you will be getting an assignment on this, then keep a copy of it. If different people monitor a magazine or two each, you will soon build up a good file with a lot of points of view and good information between you.
- Organise your files and notes so that you have a complete set of the information that you need and can look things up quickly and easily. The time spent filing things neatly will be a lot less than the time you would have spent hunting through a disorganised mess for a particular sheet of paper.

Help locally

There are quite a lot of places in most communities that can be of use to you. Here is a list of some of the main ones:

- Local libraries – there may be a town library or a library at a college or polytechnic that you can use.
- Job centres – provide a lot of leaflets on job hunting, careers, rights in employment and employment statistics.
- Citizens Advice Bureau – carries a lot of information on your rights in all sorts of situations and will also be prepared to give talks in some cases.
- Chamber of Commerce – an organisation set up by local businesses as a self-help organisation. They will have a lot of information on business in the local area and hold regular meetings. If you approach them, you will probably be welcome to attend their meetings. Many of them have a full-time local office and staff who you can approach. The local Chamber of Trade represents local traders, notably retailers, and can provide information. They are not usually so well supported as the Chamber of Commerce and don't often have a full-time office for you to approach.

- Small business centres – provide information on how to start your own business as well as all sorts of advice for small businesses.
- Banks, building societies – often have useful leaflets on the services that they provide to business and also produce leaflets on topics that might interest businesses, like exporting, cash flow or marketing.
- Local firms – depending on what work you have to do, there may be times when it would be useful for you to contact local firms for brochures, prices or information on products, or to talk to the people who work there.
- Local authority – has a lot of information about its area and will usually be willing to help you as much as it can. There are quite a lot of departments, and different councils/authorities will be organised differently. Here is a list of some of the main ones that you could find useful:

  - Surveyors department – provides information on building in the area.
  - Treasurers department – provides information on the finances of the authority including business rates.
  - Planning department – provides information on industrial development/ employment and on plans for the area, as well as on how to get planning permission.
  - Environmental health department – provides information on regulations covering health and safety.
  - Trading standards office – the experts on the law as it applies to shops and businesses. It has a lot of leaflets on consumer rights and can advise on the law in some cases.
  - Training and Enterprise Council – your local TEC can give you information on employment, industry and training in the area.

This isn't a full list of the help that you can get locally, especially in large cities where there will be more specialist sources of advice. You will also be able to find other ways of getting help and information, such as personal contacts, friends, relatives and people at work. But it does provide an outline of many of the people who are prepared to help you.

Contacting people for help

When approaching people for help, there are a few tips that will make things easier:

- Don't expect people to drop everything to deal with your request and don't pester people when you don't get information as quickly as you would like.

- If you want to contact someone, it is best to write first. This allows the person to look at what you want and deal with it at leisure. In this way, you are more likely to get a 'yes' than if you walk in off the street and ask to see the managing director.
- Try to give people time to answer questions. It's more businesslike and it gives you a far better chance of getting what you want. If they know your questions in advance, they can get the information for you or arrange a convenient time for someone to see you.
- One word of warning. If the whole class has to collect some information, try to make sure that the organisation doesn't get 24 identical requests for the information. Work out a way of sharing the information once the organisation has provided it.

Activity 39 Sources of help locally

Make a list of the local sources of help and information that aren't covered in the text and add them to the list in Table 18. Then make a note of the organisation's location and what it can tell you. Compare your list with a few others to make sure that you haven't missed anything.

TABLE 18   *Local sources of help and information*

Organisation	Address/contact person	Information
Libraries		
Job centres		
Citizens Advice Bureau		
Chamber of Trade		
Small business centres		
Local firms		
Local authority		
Chamber of Commerce and Industry		

Help further afield

There may be times when you can't get information locally and you need to go further afield. The following list gives some of the most useful organisations and their addresses:

ACAS
Advisory, Conciliation and Arbitration Service
Cleland House
Page Street
London SW1P 4ND

Information on rights in employment, industrial tribunals and settling disputes in employment.

BTEC
Central House
Upper Woburn Place
London WC1H 6HH

Information about your course and many others. BTEC provides leaflets and booklets, such as *Recognition of BTEC Awards*, which you could find useful. (Most BTEC publications should be in your college library.)

Labour Party
150 Walworth Road
London SE17 1JT

Information on their policies, as well as publicity material and leaflets. They may be prepared to answer specific questions.

Conservative Party
Smith Square
London SW1P 3JA

Information on their policies, as well as publicity material and leaflets. They may be prepared to answer specific questions.

Green Party
Freepost
London SW12 9YY

Information on their policies, as well as publicity material and leaflets. They may be prepared to answer specific questions.

Liberal Democrat Party
4 Cowley Street
London SW1P 3NB

Information on their policies, as well as publicity material and leaflets. They may be prepared to answer specific questions.

Association of British Chambers of Commerce
Sovereign House
212A Shaftesbury Avenue
London WC2H 8EW

General information and publicity material on business in the UK.

British Overseas Trade Board
1 Victoria Street
London SW1H 0ET

Information on exporting and advisory services for exporters.

Institute of Exports
World Trade Centre
London E1 9AA

Information on exporting and advisory services for exporters.

Small Firms Information Service
Freephone 2444

Advice on starting and running a small business, as well as leaflets on many topics related to small businesses.

Training Agency
Moorfoot
Sheffield S1 4BR

Information and publicity on all aspects of the training policy in the country. Note that the agency frequently changes its name – it was previously known as the Manpower Services Commission.

Equal Opportunities Commission
Overseas House
Quay Street
Manchester M3 3HN

Publicity material and advice on all aspects of equal opportunities.

Trades Union Congress
Great Russell Street
London WC1B 3LS

Information and publicity material on trade unions and the trade union movement as a whole. They often have documents relating to matters of general interest, such as employment law.

Confederation of British Industry
Centre Point
103 New Oxford Street
London WC1

Information on the economy, exporting and on major British firms. They have views, and usually documents and leaflets, on many aspects of the economy.

Bank of England
Threadneedle Street
London EC2

Publicity material on their role in the economy. Visits can be arranged.

British Tourist Authority
Queens House
64 St James Street
London SW1

Information on the tourist trade in the UK, on attracting tourists from abroad, and on events and attractions in the country.

Department of Employment
Caxton House
Tothill Street
London SW1P 9NF

Information on employment figures, training and associated matters.

Department of the Environment
2 Masham Street
London SW1

Information on plans for communications and on the environment as a whole. Planning information can also be obtained from them.

Department of Trade and Industry
1 Victoria Street
London SW1

Information on the economy, trade figures, Europe and exporting.

National Consumer Council
18 Queen Annes Gate
London SW1

Information on consumer affairs and on the law to protect them.

The European Commission
Rue de la Loi 200
1049 Bruxelles
Belgium

Information on the EEC. Visits can be arranged to the European Parliament.

European Community Information Office
8 Storeys Gate
London SW1P 3AT

Information on the EEC. Visits can be arranged to the European Parliament.

Department of Trade and Industry
1 Victoria Street
London SW1
Information on the economy, trade figures, flation and exporting

National Consumer Council
18 Queen Annes Gate
London SW1
Information on consumer affairs and on the law to protect them

The European Commission
Rue de la Loi 200
1049 Brussels
Belgium
Information on the EEC. Visits can be arranged to the European Parliament.

European Community Information Office
8 Storeys Gate
London SW1P 3AT
Information on the EEC. Visits can be arranged to the European Parliament.